Chinese Ethical and Legal Perspectives on Automated Vehicles

Chunyan Ding • Ruiping Fan
Editors

Chinese Ethical and Legal Perspectives on Automated Vehicles

palgrave
macmillan

Editors
Chunyan Ding
School of Law
City University of Hong Kong
Kowloon, Hong Kong

Ruiping Fan
Department of Public and
International Affairs
City University of Hong Kong
Kowloon, Hong Kong

ISBN 978-981-96-1276-5 ISBN 978-981-96-1277-2 (eBook)
https://doi.org/10.1007/978-981-96-1277-2

This Palgrave Macmillan imprint is published by the registered company Springer Nature Singapore Pte Ltd.
The registered company address is: 152 Beach Road, #21-01/04 Gateway East, Singapore 189721, Singapore

Acknowledgements

We express our gratitude to the Center for Public Affairs and Law at the City University of Hong Kong for generously funding our research on the Chinese ethical and legal perspectives of Automated Vehicles. Special thanks are extended to Prof. Xie Min from the Department of Systems Engineering at the City University of Hong Kong for his invaluable intellectual support in enhancing our understanding of technical issues related to automated vehicles stemming from this study. We also appreciate the exceptional research support provided by Xingjie Huang and Pei Zhi. Lastly, we would like to acknowledge the insightful comments from Springer's anonymous reviewers and the unwavering support and guidance from Jacob Dreyer, our publishing editor.

Contents

Notes on Contributors

Chunyan Ding (丁春艳) is a professor at the School of Law at the City University of Hong Kong. She holds an LLB and LLM from Peking University, an LLM from University College London, and a PhD from the University of Hong Kong. She became a qualified PRC lawyer in 2001. Before joining CityU in 2010, Ding was a Fulbright research fellow at Harvard Law School, where she conducted research on comparative health law and tort law in 2008–09. Her primary areas of research and teaching include tort law, health law, privacy and data protection, and Chinese and comparative civil law. She has authored *Medical Negligence Law in Transitional China* (2012) and published over 50 journal articles and book chapters in her research fields.

Ruiping Fan (范瑞平) is Chair Professor of Philosophy at the City University of Hong Kong. He primarily conducts bioethics and comparative philosophy research drawing on Confucian cultural resources. In addition to having over 200 journal articles and book chapters (over 100 in English and 100 in Chinese) published, he also authored *Reconstructionist Confucianism: Rethinking Morality After the West* (in English, 2010), *Contemporary Confucian Bioethics* (in Chinese, 2011), and *Contemporary Medicine and Confucian Thought* (in Chinese, 2024). He was the editor or co-editor of 15 volumes (8 in English and 7 in Chinese) regarding bioethical and Confucian studies, including three most recent volumes titled *Sex Robots: Social Impact and the Future of Human Relations* (Springer 2021), *Incentives and Disincentives for Organ*

Donation: A Multicultural Study Among Beijing, Chicago, Tehran and Hong Kong (Springer 2023), and *An East-West Dialogue on Good Governance: Learning from Each Other* (Springer 2024)..

Jianhui Li (李建会) is a professor and doctoral supervisor at Beijing Normal University, the Academic Registrar and director of Research Development and Knowledge Transfer Office at Beijing Normal-Hong Kong Baptist University. He is a former Vice Dean of the School of Philosophy, Beijing Normal University, Vice Director of the Education Management Office of Graduate School, Beijing Normal University, Distinguished Professor at the Institute of Life and Culture, Guangdong Medical University, Honorary Chairman of the Philosophy of Biology Committee of the China Society of Dialectics of Nature, Vice Chairman of the Beijing Society for the History and Sociology of Science. He mainly teaches and researches philosophy of science, ethics of science and technology, philosophy of mind and cognitive science, and the history of Chinese and foreign scientific thought. He has won several awards, including the Expert in National Medium and Long-term Strategic Planning, two Excellent Scientific and Technological Achievement Award from the Ministry of Education, one Excellent Teaching Achievement Award from Beijing, and two Excellent Achievements in Philosophy and Social Sciences from Beijing. He has completed 10 books around, such as *The Formalization of Mind and Its Challenges: The Philosophy of Cognitive Science*, and *A Study of Computationalism and Its Theoretical Problems.* He published more than 100 academic papers in journals such as *Theoretical Medicine and Bioethics*, *Journal of the History of Biology*, *Philosophical Studies*, and *Studies in Natural Dialectics.*

Yun Liu (刘云) is a research fellow at the Tsinghua University Think Tank Center, and assistant director of the Institute for Studies on Artificial Intelligence and Law at Tsinghua University. His research primarily focuses on issues at the intersection of law and information technology, including computational legal theory, data governance, artificial intelligence governance, and cybersecurity. He was a visiting scholar at Ohio State University Law School from August 2016 to August 2017. In 2017, he obtained his doctoral degree in law from China University of Political Science and Law. He has published more than 20 academic papers in mainstream academic journals. He has undertaken multiple research projects as a key research member, including "Research on the Implementation of the New Standardization Law", "Research on Intelligent Assistive Technology in the Context of Judicial Processes Involving Concerned Civil and Commercial Cases", and "Research on the Legal Governance of the Internet Economy".

Weixing Shen (申卫星) is a professor and PhD supervisor at the Law School of Tsinghua University, Director of the Institute of Artificial Intelligence and Law of Tsinghua University, Executive Director of China Law Society, Vice Chairman of the China Law Society Network and Information Law Association, and Vice Chairman of China Law Education Society. He conducts research in civil law, health law, and computational law. He has published 19 monographs and teaching materials, including the Research on the Basic Categories of Civil Law, the Principles of Property Law, and the Research on the Basic Theory of Expectation Right, and published more than 100 legal academic papers in such publications as the *Chinese Social Science*, *Chinese Journal of Law*, and *China Legal Science*. He also chairs the Research on Intelligent Auxiliary Technologies for Trial of Hot Cases and Cases of People's Livelihood, a national key R&D project undertaken by the Ministry of Science and Technology, and the Research on Legal Guarantee for the Internet Economy, a major project funded by the State Council for Social Science and Technology. He has also undertaken such projects as the Research on Key Issues Concerning Network Data Security Responsibilities of the Office of the Central Leading Group for Cyberspace Affairs, and the Research on Major Legislation Issues Concerning Artificial Intelligence by the Ministry of Science and Technology. He edited and published the Research on Digital Economy and Cyber Law.

Jue Wang (王珏) is a professor in the Department of Philosophy at Xi'an Jiaotong University. Her main research fields are foreign philosophy, comparative philosophy, and applied ethics. She is a member of the Editorial Board of the *International Journal of Chinese and Comparative Philosophy of Medicine* and a member of the International Cooperation and Exchange Group, Medical Ethics Branch Committee of the Chinese Medical Association. She has received several awards, including the National Social Science Research Grant, the Humanities and Social Sciences Research Grant, and the Award for Visiting Scholarship. She has published several articles and book chapters, including "A Confucian Reflection on Transhumanism: How to Regulate Our Posthuman Future", "Should We Develop Empathy for Social Robots", and "The Issue of Intergenerational Equity in the Context of Population Aging: From the Perspective of

Confucian Ethics". Currently, her research focuses on the phenomenology of the body, the philosophy of artificial intelligence, and the ethics of technology.

Ying Wang (王莹) is a professor at the Digital Law Teaching and Research Center at the Law School of Renmin University of China and Vice Director of The Law and Technology Institute of Renmin University. Her research has been focused on cybercrime and cybersecurity, AI, and data law. She became a visiting scholar at the Institute for Artificial Intelligence, University of Würzburg, in August 2018, and a research fellow at the Institute for Digital Translation, Bavaria, in February 2021. Currently she is a visiting scholar at the Law School of Columbia University. Her key achievements include the publication of a series of monographs and articles, as well as leading multiple research projects funded by the National Social Sciences Foundation and major IT and AI companies. Two of her notable monographs include *Chinese Criminal Jurisprudence: Experience, Reflection and Construction* (in Chinese) and *Protection of Copyright in Criminal Law: A Comparative Study of German and Chinese Criminal Law* (in German). Additionally, she has published influential articles on algorithm liability and the regulation of automated driving in leading Chinese journals, drawing on her active participation in the legislative and rule-making processes in these areas.

Qian Wang (王前) is a professor at the Department of Philosophy at the Dalian University of Technology. His research mainly focuses on the philosophy of technology, engineering ethics, science and technology ethics. He has published over 200 journal articles, monographs, and book chapters. He was awarded the Chinese Academic Degree and Postgraduate Education Award, Postgraduate Teaching Quality Excellence Award, and Dalian Outstanding Expert Award. He has conducted several research projects, including the value selection and risk prediction of ethical governance of science and technology, as well as the theory and application of responsible innovation.

Shilig Xiao (肖石灵) is Lecturer in Law at the University of Manchester. He was previously a post-doctoral researcher at the School of Law at the City University of Hong Kong. His research interests lie in the following areas: constitutional review, constitutional orders of authoritarian and hybrid regimes, law and politics, legislative studies, judicial review, administrative rule-making, and comparative public law. His publications have been in the *International Journal of Constitutional Law*, *Wisconsin International Law Journal*, *Journal of Comparative Law*, *Asia Pacific Law Review*, and *Hong Kong Law Journal*. His monograph *Judicial Review in Greater China* will be published by Routledge in 2025.

Katharina Yu is an associate professor (administration) at the Beijing Normal-Hong Kong Baptist University, where she also serves as the Director of the International Development Office (IDO). She holds a PhD in Indology from Ludwig Maximilian University of Munich (LMU München), where she conducted research on the Prestige of Asceticism in Ancient India. Prior to her doctoral studies, she studied Indology, Sinology, and Indo-European Linguistics at the University of Muenster. Her research interests span across asceticism, philosophy, philology, cultural studies, and intercultural competence.

Wei Zhang (张卫) is a professor in the Department of Philosophy at Dalian University of Technology and a doctoral supervisor in Philosophy. He currently serves as the Vice Director of the Science, Technology and Engineering Ethics Professional Committee of the Chinese Society for Dialectics of Nature, Executive Member of the Science and Technology Ethics Professional Committee of the Chinese Ethics Society, member of the Science and Technology Culture Professional Committee of the Chinese Society for Dialectics of Nature, and a member of the Technical Philosophy Professional Committee of the Chinese Society for Dialectics of Nature. He also serves as a reviewer for several professional journals both domestically and internationally. He was awarded the "Ministry of Education's Academic Newcomer Award for PhD Students". His research mainly focuses on the philosophy of technology and technology ethics. He leads the National Social Science Fund Youth Project, "Research on the Theoretical Reconstruction of Technoethics Based on Intrinsicism", and the China Postdoctoral Science Fund General Project, "Research on the Moral Agency of Artifacts". He has published over 40 papers in core journals at home and abroad, including *Philosophical Trends*, *Studies in Dialectics of Nature*, *Studies in Philosophy of Science and Technology*, *Ethical Studies*, *Morality and Civilization*, *Social Epistemology*, *Techne*, and *Technology in Society*, among others. Several of his articles have been reprinted in full in the *People's University Digest*. He has also published the monograph *Research on Intrinsicist Technoethics*.

List of Figures

List of Tables

PART I

Introduction

CHAPTER 1

Introduction: Exploring Chinese Ethical and Legal Perspectives on Automated Vehicles

Chunyan Ding and Ruiping Fan

Introduction

This volume presents Chinese ethical and legal viewpoints on the moral and regulatory concerns surrounding the deployment of automated vehicles (AVs). Notably, China is at the forefront of autonomous driving road testing, experiencing rapid advancements in this field. Major metropolitan

We wish to express our gratitude to Xingjie Huang for her valuable assistance in helping us complete this chapter.

C. Ding (✉)
School of Law, City University of Hong Kong, Kowloon, Hong Kong
e-mail: chunding@cityu.edu.hk

R. Fan
Department of Public and International Affairs, City University of Hong Kong, Kowloon, Hong Kong
e-mail: safan@cityu.edu.hk

C. Ding, R. Fan (eds.), *Chinese Ethical and Legal Perspectives on Automated Vehicles*,
https://doi.org/10.1007/978-981-96-1277-2_1

cities in China, including Shenzhen, Guangzhou, Beijing, Shanghai, and Wuhan, have actively engaged in such testing. Specifically, Wuhan has emerged as a notable case, offering valuable background information for us to reflect on the crucial ethical and legal matters related to the adoption of AVs in China. For example, data shows that by the end of 2023, the total open road mileage for autonomous driving tests in Wuhan has surpassed 3378.73 kilometers, covering 12 administrative regions, with a radiating area of approximately 3000 square kilometers and reaching a population of over 7.7 million. The total mileage and the number of open areas in Wuhan rank first in the country, making it the largest autonomous driving service area in the world (Zhou & An, 2024).

Indeed, Wuhan is also a demonstration area in China that issued commercial licenses for autonomous driving earliest. In August 2022, Wuhan became one of the first cities nationwide to issue licenses for remote driving tests. Baidu's autonomous driving service platform, "Apollo Go", obtained the first license for remote driving tests, marking the entry of Wuhan's autonomous driving applications into a "fully unmanned phase." From August 2022 to August 2023, Baidu's "Apollo Go" deployed 300 unmanned vehicles in Wuhan, with operational areas and covered population experiencing exponential growth. On August 25, 2023, Baidu's "Apollo Go" launched shuttle services at Wuhan Tianhe Airport, making Wuhan the first city in China to implement autonomous driving shuttle services at an airport (Xie et al., 2023). On February 27, 2024, Baidu's "Apollo Go" fully unmanned autonomous vehicles drove across the Yangsigang Yangtze River Bridge and the Baishazhou Yangtze River Bridge in Wuhan, making Wuhan the first city in the country to achieve the demonstration operation of intelligent connected vehicles crossing the Yangtze River (Xun & Gong, 2024).

Wuhan City is currently formulating relevant legal regulations, with its draft already entering the review stage of the Municipal People's Congress Standing Committee. Among them, Article 14 encourages and supports market entities to participate in the investment, construction, operation, and maintenance of intelligent connected vehicle infrastructure. Article 17 stipulates the establishment of the city's intelligent connected vehicle safety supervision platform and data center based on the Vehicle-City-Internet platform to avoid redundant investments. Article 36 proposes the establishment of an inclusive and prudent regulatory framework and an innovation and entrepreneurship fault-tolerant and immunity mechanism to promote the development of intelligent connected vehicles. Article 40 specifies the intensification of efforts in protecting intellectual property rights. The "Regulations (Draft)" also focus on security assurance. Articles 26 and 27 stipulate that the municipal and district people's governments should establish a sound intelligent connected vehicle security assurance system and strengthen safety supervision of intelligent connected vehicles. Article 28 specifies the responsibilities and obligations of relevant entities in the intelligent connected vehicle ecosystem, including producers and sellers of intelligent connected vehicle products. Article 30 mandates the adoption of technical measures and other necessary steps to ensure data and personal information security (Xiao, 2024). The experts emphasize that legislation should follow some basic principles, with the primary one being the "absolute subjectivity of humans and the principle of human safety first" (Xun, 2024).

However, general principles must be clearly defined and specified to be substantiated and effective. In fact, with the widespread popularity of "Apollo Go" in Wuhan, controversies about whether it will replace drivers are constantly heard. In response, the person in charge of the Wuhan Municipal Transportation Bureau told the media that the (rental) car industry in Wuhan is relatively stable at present, despite numerous online rumors. Additionally, the safety of "Apollo Go" has recently come under scrutiny. On the afternoon of July 7th, 2024, "Apollo Go" collided with an electric vehicle running a red light at a traffic light intersection in Wuhan, causing some congestion. Prior to this incident, the "Wuhan City Message Board" received multiple complaints about "Apollo Go," such as vehicles stalling at green lights, running into the middle of intersections at red lights, and experiencing difficulties when turning on the road (Liu, 2024). The Wuhan authorities did not halt Baidu's testing due to this incident but stated, "Feedback is being provided to the relevant departments, recommending optimization of the unmanned driving system to ensure good traffic order on the roads" (Tang, 2024).

The case of Wuhan underscores the imperative for a thorough examination of the ethical and legal considerations intertwined with the deployment of AVs in China. These issues extend beyond concerns like safety and unemployment, although undeniably important. They intersect with the overarching Chinese developmental strategy concerning the human-machine relationship. Delving into topics such as the distinction between algorithmic morality and human morality, the programming of software in AVs to navigate emergent ethical dilemmas, safeguarding privacy, regulating associated risks within legal frameworks, defining civil liability through legislation, and establishing effective data governance mechanisms becomes paramount. Wuhan's example serves as a compelling reminder that a nuanced understanding of these complex issues is essential for the responsible advancement of AV technology in China. That is why we have strategized and coordinated this book project for in-depth examinations of these challenges.

There is a significant body of literature addressing the ethical and regulatory considerations surrounding AVs. While there are shared moral principles across cultures in these studies, there are also disagreements among researchers. For instance, it is widely acknowledged that establishing clear guidelines for responsible on-road testing is crucial to mitigate risks to road users (European Commission, 2020). Furthermore, ethical trajectory planning algorithms for autonomous vehicles should prioritize equitable risk distribution among road users, adhere to the principle of minimizing overall risk, opt for trajectories with the least overall risk, and prioritize the safety of all road users (Geisslinger et al., 2023). Undoubtedly, the significance of ethical considerations in the decision-making processes of autonomous systems cannot be overstated, ensuring that these algorithms are ethically sound to guarantee robust and secure decision-making (Xiao, 2024). However, while some argue for upholding individuals' rights to control their fate amidst advancing AI technologies (Porter et al., 2024), others propose making robotic driving more analogous to human driving and vice versa (Nyholm & Smids, 2020). Regarding ethically sound resolutions to trolley-dilemma-like scenarios in AV deployment, some advocate for a mandatory ethics framework instead of allowing individual drivers to select personal ethics settings to serve the societal interest by minimizing harm risks in traffic (Gogoll & Müller, 2017). In contrast, an integrative ethical decision-making model encompassing multiple interconnected stages has been suggested to elucidate the process of pluralistic ethical decision-making (Rhim et al., 2021). Additionally, arguments have been made that AVs should never be programmed to swerve in such situations to preserve specific areas like sidewalks and pedestrian crossings as inherently secure spaces (Lawlor, 2022).

In our view, previous research has often lacked culturally relevant investigations, which is a shortcoming since some, if not all, essential ethical and legal considerations related to AVs are not universally applicable across cultures. This book project aims to address this gap by offering Chinese culture-specific ethical and legal viewpoints on AV development within Chinese society. While existing literature notes the importance of cultural factors in ethical and legal decision-making within AV design, many studies have primarily centered on general psychological differences rather than delving into the specific cultural nuances deeply rooted in particular societies. While these inquiries are valuable, they frequently neglect the normative aspects of culture, potentially providing only superficial insights.

In Section II of this introduction, we elucidate the necessity of conducting culturally relevant studies on the ethical and legal aspects of AVs. This section articulates the rationale behind this book project's aim to integrate normative inquiries with quantitative studies to explore fitting Chinese ethical and legal perspectives on the design and regulation of AVs within Chinese society. In Section III, we provide a synopsis of each chapter. Lastly, Section IV outlines the significance and limitations of this book project and directs attention towards potential avenues for further research.

Culture-Relevant Studies

While the ethical and legal challenges surrounding AVs possess universal characteristics, they are also deeply intertwined with specific cultural contexts. Therefore, conducting culturally sensitive studies on these matters becomes imperative. Examining these issues in the Chinese context, for instance, requires a nuanced understanding of Chinese culture and societal norms. China's rich cultural heritage and unique societal structures influence how these technologies are perceived, adopted, and regulated within the country. Factors such as the emphasis on the family, the role of technology in shaping social interactions, and the government's approach to governance all play a crucial role in shaping the ethical and legal frameworks surrounding AVs in China. By delving into these cultural nuances, researchers can better comprehend the intricacies of these issues and tailor solutions that align with the values and expectations of Chinese society.

Research focusing on the cultural implications of ethical and legal considerations surrounding the deployment of AVs has not been prominently featured in existing literature. However, this does not imply a lack of evidence supporting the importance of cultural factors in ethical decision-making. For instance, earlier research conducted on the renowned "Moral Machine" platform developed by the MIT lab has shed light on the cultural relevance of people's moral choices when programming the software of AVs. This platform, aimed at gathering human perspectives on moral judgments by artificial intelligence, specifically AVs, presents moral dilemmas where a driverless car must select the lesser of two evils, such as deciding between saving two passengers or five pedestrians. Participants are prompted to evaluate which outcome they find more acceptable as outside observers, comparing responses and allowing them to gauge their perspectives against those of others (MIT, 2024). By 2018, the platform had hosted online ethical decision-making exercises in ten languages, accumulating 40 million decisions from respondents across 233 countries. A formal publication by the researchers detailed three key findings: most people tend to prioritize human lives over animals, to save a greater number of lives over a smaller number, and to prioritize the lives of younger individuals over older ones (Awad et al., 2018). Notably, significant cultural variations were observed among these findings, with a lower preference for preserving young lives over older ones in Eastern countries influenced by Confucian and Islamic values, contrasting with a higher preference in Southern countries, including those in Latin America (Awad et al., 2018, p. 61).

Regrettably, these researchers did not further investigate these cultural distinctions in terms of the substantive content of cultural norms and values deeply rooted in specific societies. Instead, they shifted focus towards offering broad explanations of cultural variations through the lens of the psychological concept of "relational mobility." This shift involved analyzing a dataset derived from their Moral Machine platform, encompassing moral preferences expressed by 70,000 participants across 10 languages and over 40 countries up to March 2019. In psychology, relational mobility pertains to the ease with which individuals can form new relationships. Their findings indicate that when making decisions concerning the classic Switch, Loop, and Footbridge scenarios of the trolley problem, regions with low relational mobility—where individuals are more cautious about potentially alienating their existing social connections—tend to resist making sacrifices for the collective good, particularly observed in Eastern countries. This reluctance may be attributed to the signal conveyed by such refusals. Notably, the United States stands out among Western nations for its higher propensity to endorse sacrifice, whereas China and Japan rank among the Eastern countries least inclined to support sacrificial actions (Awad et al., 2020).

Such a common trend of psychological generalization and a lack of normative specificity can be observed in various cross-cultural research within the field.[1] While such studies and their findings play a useful role in highlighting cultural disparities concerning ethical decision-making and legal frameworks in the design and use of AVs, they also exhibit notable limitations. Primarily, these studies tend to concentrate solely on quantitative analyses rather than delving into qualitative investigations. They present statistical outcomes of individuals' choices without delving into the underlying considerations and rationales guiding these decisions, resulting in only superficial findings. Furthermore, they often overlook the substantive frameworks and intricacies of mainstream cultures (such as Confucian moral culture) within broad national and societal contexts (such as many regions of China), opting instead to present a balanced overview of multiple specific cultures. This broad approach to research tends to overlook the fundamental function of the predominant moral culture shaping behaviors within a society. Additionally, these studies disregard the essential role of normative ethical theories and principles ingrained in specific cultures, which stand cardinal positions in influencing individuals' ethical perspectives and decisions regarding the deployment of AVs. By neglecting the normative aspect of culture, there is a risk of adopting a deterministic perspective on individual choices and decisions, viewing individuals as products of their societal conditions and cultural backgrounds rather than as autonomous agents acting based on their voluntarily embraced reasons and values to respond to the ethical and legal challenges posed by the utilization of AVs.

[1] For example, in a study exploring the cultural impact of control sharing in automated driving and examining cross-cultural perspectives on control allocation in ethical decision-making involving human drivers and AI-enabled automated vehicles, researchers analyzed a total of 771 participants from the United States, India, and Nigeria. The findings revealed that individuals from individualistic cultures, such as the United States, exhibited a preference for greater control over AI in ethical decision-making compared to participants from collectivist cultures like India and Nigeria. The researchers argued that this observation supported the hypothesis that individualism and lower power distance correlate with a desire for increased control over AI in ethical decision-making (Wan et al., 2022). In another online experimental study conducted across China, Germany, Japan, and the United States, focusing on four overtaking scenarios where automated vehicles made decisions regarding overtaking or remaining in lane and the potential hindrance level to other drivers, it was found that cultural influences play a significant role in the acceptance of overtaking maneuvers by automated vehicles. Specifically, the study highlighted that Japanese users tend to prefer staying in lane regardless of hindrance level, Chinese drivers show a higher acceptance for overtaking over remaining in lane, and German and US participants lean towards allowing the automated vehicle to overtake when hindrance is minimal (Edelmann et al., 2021).

This book project endeavors to integrate normative inquiries and quantitative studies to delve into appropriate Chinese ethical and legal standpoints concerning the design and governance of AVs within Chinese society. As widely recognized, Confucianism and Daoism stand as the predominant Chinese cultural legacies that continue to significantly influence Chinese lifestyles and decision-making processes. These traditions should also be factored in when determining the best approach for developing a fitting Chinese strategy for AV advancement and implementing suitable regulations for their operation. The fundamental concepts and principles embedded in these traditions warrant scrutiny to illuminate how the Chinese populace should address the ethical dilemmas brought about by AVs, including decisions during brake failures and safeguarding privacy. By incorporating these cultural perspectives, this project aims to offer comprehensive insights into how China can navigate the ethical complexities associated with AVs, paving the way for ethically sound and culturally resonant approaches to AV development and regulation.

When it comes to the legal regulations and liability framework for AVs, China must carefully weigh and balance a variety of competing interests within its unique socio-legal and cultural context. On one hand, as a proponent of high technology and an advocate for the advancement of the automated vehicle industry and market due to the potential long-term social benefits they offer, Chinese law is tasked with providing clear guidance and detailed rules for the phases of AV road testing, demonstration deployment, and commercial deployment. Simultaneously, given that connected AVs collect and operate on vast amounts of users' personal information, vehicle operational data, and road environment data, the Chinese government must establish a comprehensive data governance system to rigorously oversee and supervise privacy and security concerns stemming from AV usage. On the other hand, rooted in the traditional Chinese perception of collective rights and responsibilities, Chinese law tends to lean towards considering an industry-based collective liability framework to address civil liability in AV accidents as well as the corresponding liability insurance regimes. This approach aims to foster user confidence and trust in the future of transportation, particularly in highly automated vehicles.

In conjunction with normative studies, we conducted a survey to gauge Chinese preferences regarding the resolution of ethical dilemmas and legal issues surrounding AVs in two key cities, Hong Kong and Shenzhen. The surveys were administered through online panels hosted by a globally respected data collection firm. A total of 1000 participants were randomly selected to respond to our questionnaires, with 500 individuals hailing from Hong Kong and another 500 from Shenzhen in mainland China. The choice of Hong Kong and Shenzhen as survey locations was deliberate, aimed at capturing a diverse array of perspectives within the Chinese cultural milieu. Despite their geographical proximity, these cities boast distinct characteristics in terms of culture, economy, legal frameworks, and traditions, rendering them as valuable representatives of broader Chinese viewpoints. Moreover, Shenzhen's tech-forward approach contrasted with Hong Kong's more conservative stance. The 1000 participants were tasked with answering 16 ethical queries and 5 legal questions. By analyzing these responses alongside the normative ethical considerations outlined in preceding chapters, readers are presented with a comprehensive and well-rounded insight into Chinese ethical and legal viewpoints.

Summary of Chapters

This book is structured into four parts. Following this introductory section, the second part of the book comprises four chapters that delve into the ethical complexities surrounding AVs. Chapter two in this segment explores a Chinese AV strategy through a Daoist lens. Daoism, a time-honored Chinese tradition revolving around the concept of Dao (way, 道), elucidates the fundamental principles governing the universe and directing human endeavors. According to Qian Wang and Wei Zhang, the Daoist notion of "Dao" epitomizes an ideal state in human technical operations where there exists perfect harmony among the operator, the object being operated on, and the tools utilized. They draw upon the Daoist parable of "Butcher Ding Carving an Ox" to elucidate the Daoist tenets of "self-momentum" (ziran，自然) and "non-action" (wuwei，无为), underscoring the significance of aligning with the inherent characteristics of entities and eschewing coercive control. In their perspective, Daoism rejects unnatural technology that leads to resistance, rebound, and detrimental conflict, advocating for a naturalistic approach that collaborates with the environment and adheres to its intrinsic drives to realize "governing technology by Dao" in the advancement of AVs.

To achieve this objective, the authors posit that it is imperative to take into account the concerns of all stakeholders involved with AVs, encompassing designers, manufacturers, marketers, managers, users, scholars, and the general public. Daoism accentuates the necessity of transcending the dichotomy between machines and humans, striving instead for a harmonious coexistence among diverse technical components. Referencing the tale of an elderly man declining a well-sweep, the chapter symbolizes the Daoist critique of "mechanical affairs" and a "mechanical mindset," denoting the adverse effects of machinery on individuals' physical and mental well-being and the encroachment of "mechanical affairs" on human lives. The authors advocate three guiding principles: the "Human-Centeredness" principle (*yi ren wei ben*, 以人为本), the "Perceiving with the Heart" principle (*yi xin guan wu*, 以心观物), and the "Virtue and Benefit Alignment" principle (*de fu yi zhi*, 德福一致). They posit that the core challenge posed by AVs lies in harmonizing their development with the preservation of alternative modes of transportation, alongside addressing specific ethical concerns like safety hazards, data privacy and accountability, human alienation, and moral dilemmas. Emphasizing the alignment of AV deployment with Daoist social and ecological values, the authors advocate for the pursuit of "governing technology by Dao" within the realm of AVs.

Chapter three, authored by Jue Wang, shifts the focus to the Confucian Dao concerning AVs. Wang examines the ethical dimensions surrounding algorithmic morality within AVs and advocates for a Confucian relational perspective on machine ethics as an alternate ethical framework. She contends that the confined scope of trolley dilemmas falls short in addressing the moral quandaries inherent in AVs, emphasizing the necessity for a comprehensive moral landscape that considers the diverse interests and stances of all involved parties. According to Wang, Confucian ethics offers valuable insights for navigating moral dilemmas, offering a profound moral structure beyond conflicting ethical doctrines. She critiques utilitarian and deontological approaches to algorithmic ethics for oversimplifying ethical decision-making into a matter of mechanical normative choice, thereby obscuring the deeper inquiry of whether machines should supplant humans in decision-making processes.

Wang advocates for a Confucian relational outlook on machine ethics as a means to establish a holistic and equitable ethical framework for tackling the intricate ethical hurdles posed by AVs. Central to this approach is the recognition of the distinct roles of humans and machines in ethical decision-making. Machines are portrayed as implicit ethical agents designed to serve human objectives and subject to human oversight. The cardinal virtue of a machine lies in its allegiance to its designated task, with AVs being unequivocally barred from making decisions that jeopardize the safety of their passengers. Furthermore, the concept of Confucian harmony (he, 和) epitomizes the ultimate order of the universe, where all entities move and find their purpose without contradiction. This paradigm fosters a domain within the world where humans and machines can coexist and engage in a harmonious and prosperous manner. Nonetheless, as Wang underscores, when applying the Confucian notion of harmony to address AV-related issues, it is crucial to discern the fundamental disparity between humans and machines (such as AVs) concerning moral agency. Machines should never be entrusted with the authority to make moral judgments in lieu of humans.

Chapter four, titled "I must have a Steering Wheel in My Automated Car to Protect My Family," penned by Ruiping Fan, introduces a culturally pertinent Confucian approach to managing brake failure incidents in AVs. Specifically, this chapter delves into the impact of Confucianism as the predominant moral ethos in many East Asian regions, aiming to ensure the appropriateness of addressing these challenges through a Confucian virtue ethical lens. Fan scrutinizes the constraints of various comprehensive universalist frameworks in AV ethics—such as consequentialist and self-protective paradigms, deontological approaches, and virtue ethics—underscoring their inadequacies and incongruities with the moral ethos prevalent in Confucian-influenced regions. Additionally, it offers a succinct overview of the fundamental tenets of Confucian virtue ethics, which form the bedrock of moral philosophy in these areas, while also elucidating the challenges associated with expecting diverse cultural regions to adopt identical AV algorithms for addressing brake failure scenarios.

Drawing upon Confucian family-centric virtue ethical standards, this chapter puts forth specific recommendations for programming AVs in instances of brake failure, advocating for tailored programming based on the distinct purposes of various types of AVs. Fan contends that automated trucks tasked with transporting goods and animals should be programmed to prioritize swerving towards a solid barrier to shield bystanders. Conversely, automated buses ferrying numerous passengers ought to be programmed to prioritize maintaining their trajectory to ensure the safety of occupants. Regarding automated small cars designed for individual or family use, the programming should allow for swerving towards a solid barrier to protect bystanders. However, the chapter underscores the necessity for each of these small cars to retain a steering wheel, affording human drivers the option to intervene if necessary. This provision empowers individuals to exhibit virtuous behavior and make sound decisions to safeguard their family members amidst transportation emergencies, drawing on the family-oriented ethical concerns embedded within Confucian moral teachings.

The final chapter of the second part addresses ethical considerations concerning the safeguarding of individual privacy in the utilization of AVs. Jianhui Li and Katharina Yu undertake a comparative analysis between Western ethical principles and those rooted in Confucianism regarding the preservation of personal privacy. They highlight that contemporary Western societies predominantly address privacy through the lens of individual rights. Various scholars have contributed to the Western conceptualization of privacy. Some define privacy as "the right to be left alone," advocating for legal safeguards to uphold personal autonomy and individual liberty. Others introduce a four-dimensional privacy framework encompassing solitude, intimacy, anonymity, and reserve. Alternatively, the "restricted access/limited control (RALC) theory" posits that privacy hinges not solely on absolute control over data access but on establishing normative boundaries to prevent unauthorized intrusion.

In contrast, the authors delineate a Confucian stance on privacy protection that underscores the importance of interpersonal relationships, societal equilibrium, and ethical behavior. While Confucianism does not present a formalized privacy doctrine, it incorporates various principles aimed at preserving personal privacy intricately tied to the concept of "*li*" (ritual or rite, 礼). This concept underscores the significance of appropriate decorum in interpersonal engagements, such as announcing one's presence before entering a space, showing respect through lowered gaze, and refraining from unwarranted scrutiny. The Confucian Book of Rites offers examples of rituals indirectly safeguarding privacy by delineating personal boundaries and honoring individual intentions. These ritualistic norms not only establish Confucian etiquette in social interactions but also extend to the protection of personal information, emphasizing the importance of refraining from public disclosure of private details.

By juxtaposing Western and Confucian perspectives on privacy theories, the authors argue that both traditions acknowledge the significance of privacy protection. While Western ideologies center on individual rights and legal structures, Confucianism places emphasis on social cohesion, interpersonal connections, and ethical comportment. Both paradigms provide valuable insights for upholding privacy in the realm of AVs.

Western principles like informed consent, purpose limitation, data minimization, transparency, and accountability can be implemented in AV technologies to safeguard individual privacy rights. Concurrently, Confucian ethical directives, including fostering trust, transparency, collective consent, and anonymized data processing, can further fortify privacy safeguards. By amalgamating Western and Confucian viewpoints, a holistic approach to privacy preservation in the age of AVs can be realized.

Part three of this book presents Chinese legal perspectives in three chapters. Chapter six, authored by Ying Wang, delves into the evolving risks posed by automated vehicles and explores the regulatory strategies being considered by various representative jurisdictions to address these legal challenges. The chapter recognizes the transformative potential of automated driving technology in enhancing transportation safety and efficiency, while also highlighting the emergence of new and intricate risks associated with automated vehicles. These risks include concerns related to data privacy and security, algorithmic harm, and the complexities of assigning responsibility due to the involvement of multiple stakeholders. The chapter conducts a comparative analysis of regulatory frameworks for automated driving across different countries. For instance, China's comprehensive regulatory approach encompasses road testing, demonstration deployment, and commercial deployment phases. The UK has established a robust code of practice for testing automated vehicles and is progressing towards developing a comprehensive liability framework. In the US, the National Highway Traffic Safety Administration (NHTSA) has launched initiatives to enhance transparency and safety in automated vehicle testing, although specific performance measures and testing standards are yet to be established. Germany's approach centers on ensuring technical safety and addressing liability issues.

The author highlights the critical need to tackle the human-machine interface challenge as automated driving systems advance in sophistication. She cautions against the dangers of complacency and excessive dependence on automation, which can result in accidents. The intricate matter of liability in accidents involving automated vehicles demands meticulous examination of the legal standing of automated driving systems and the distribution of responsibility among manufacturers, operators, and users. She stresses the importance of establishing a comprehensive and flexible regulatory framework that can adapt to the swift progress in automated driving technology. Furthermore, she advocates for international cooperation and alignment of regulations to guarantee the safe and accountable introduction of automated vehicles on a global scale.

The upcoming Chapter seven, penned by Chunyan Ding, is titled "Exploring an Appropriate Legal Framework for Civil Liability in Automated Vehicle Accidents." Its objective is to delve into the development of a fitting nationwide legal structure for civil liability in automated vehicle accidents in China, striking a balance between enhancing safety and fostering the adoption of automated vehicles. While there are no specific statutory regulations governing civil liability for automated vehicle accidents at the national level, the chapter scrutinizes relevant rules issued by various departments of the State Council and local regulations implemented by major cities to offer guidance on the road testing, operation, and civil liability aspects of automated vehicles. It puts forth recommendations for amending the existing road accident law, product liability law, and liability insurance schemes to tackle the legal complexities stemming from automated vehicles. The chapter offers timely insights for potential legislative reforms in China in this domain.

The author classifies automated vehicles into two categories: low-level automated cars and highly automated cars, to differentiate the civil liability implications of accidents involving these distinct vehicle types. In the case of low-level automated cars, where the user-in-charge are intermittently required to assume control, they are expected to possess valid driving credentials, comply with road safety regulations, and uphold a duty of care towards both fellow road users and passengers. In instances of negligence, the user-in-charge is held responsible for providing compensation. To address information disparities and evidentiary obstacles encountered by victims, the author suggests transferring the burden of proof regarding breach and causation issues from the victim to the user-in-charge. The author also proposes the installation of a "black box" in automated vehicles to aid accident investigations.

In cases where driverless highly automated vehicles are implicated in accidents stemming from hardware, software, or information flaws, the author proposes that product liability serves as an appropriate legal avenue for victims seeking redress. By reexamining the existing Chinese product liability law, the author recommends redefining the definitions of "manufacturer" and "defect" within the realm of automated vehicles. This entails imposing joint and several liability on multiple manufacturers and sellers for product defects, mirroring a "common enterprise liability" framework. Furthermore, such industry-based strict liability could also encompass business controllers and owners of automated vehicles utilized for commercial endeavors. The author recommends shifting the burden of proof on product defects and causation to the defendants and argues against applying the state-of-the-art defense to defective automated vehicles, given their reliance on performance during the road-testing phase. Lastly, the author stresses the necessity of establishing a comprehensive liability insurance framework to effectively distribute the legal risks associated with accidents involving automated vehicles.

The subsequent chapter in this section turns to the discourse on data governance within the realm of connected automated vehicles, co-authored by Weixing Shen and Yun Liu. Unlike traditional vehicles, the functioning of automated vehicles heavily relies on vast amounts of road environment data and the ongoing recording and transmission of data. This chapter underscores the escalating data gathering from autonomous vehicles and the governance challenges that various jurisdictions are grappling with, referencing recent regulatory advancements in the US, the EU, and China. The continuous collection and sharing of data give rise to privacy risks, including unauthorized access to sensitive information, potential data misuse, and concerns regarding data security. Recent incidents involving data breaches from vehicle sensors underscore these vulnerabilities.

By examining the diverse types of data integral to the functioning of automated vehicles, such as in-vehicle occupant data, vehicle operational data, and road environment data, the authors comparatively delineate the varying regulations across different jurisdictions that target data privacy and security in connected vehicles. Specifically, the regulations in the US prioritize the security of information and communication technology, while the EU guidelines focus on personal data processing in connected vehicles. Chinese provisions mandate the submission of annual data security reports and outline specific data collection and storage requirements. The authors advocate for the establishment of an international framework for data governance within the realm of autonomous vehicles. This framework should be designed to accommodate the intricacies of cross-border data flow, safeguard personal information, and facilitate technological progress in autonomous driving.

This transitions the book into its final section, featuring a chapter authored by Chunyan Ding, Ruiping Fan, and Shiling Xiao, presenting the outcomes of a 2023 survey conducted in Hong Kong and Shenzhen. The survey aimed to capture the inclinations of Chinese individuals concerning AV design, ethical considerations, and regulatory frameworks, examining the repercussions of AVs on Chinese communities through a selection of randomly chosen participants in Hong Kong and Shenzhen. This marks the inaugural report in literature on a large-scale survey concerning Chinese preferences regarding AV development.

The authors note that the survey has unveiled intriguing disparities between Chinese and Western perspectives on certain ethical considerations. For instance, akin to Western viewpoints, 76.2% of the respondents in this survey endorsed the notion that AVs should prioritize the safety of their drivers and passengers over other road users in potential brake failure scenarios. However, regarding the willingness to acquire an AV programmed to prioritize saving pedestrians over occupants by swerving onto a solid barrier during brake failure, responses diverged significantly from Western stances. Notably, 77.6% of the participants expressed their readiness to purchase such an AV (with marginal variance between Hong Kong and Shenzhen), contradicting to their prementioned preference. The authors attribute these contrasting choices to specific Chinese cultural influences. While there is a general preference for AVs prioritizing internal users' safety in emergencies, participants seem open to purchasing AVs programmed to protect external road users, possibly due to the influence of Confucian meritocratic principles. Individuals may trust decisions made by skilled and meritorious figures, such as AV experts, advocating for prioritizing external user safety in crisis situations.

Interest in AVs surged as automation levels advanced, with Shenzhen respondents displaying keener interest across various automation levels compared to their Hong Kong counterparts. Both groups believed that AVs would positively impact society, although Hong Kong participants exhibited slightly less optimism than those from Shenzhen. A significant majority (75.5%) of the participants, with no notable difference between Hong Kong and Shenzhen, concurred that a compact AV intended for personal or familial use should be programmed to swerve onto a solid barrier rather than target pedestrians in a crisis, prioritizing the lives of multiple pedestrians. Moreover, 87.1% of the participants (89.2% in Shenzhen and 85% in Hong Kong) advocated for integrating a steering wheel in AVs to enable human intervention and decision-making at any point. This widespread consensus, as perceived by the authors, underlines the Chinese populace's belief that retaining a steering wheel empowers them to safeguard their family members' lives within an AV based on their own discretion.

The authors contend that these findings can offer valuable insights to automobile manufacturers, policymakers, and ethicists regarding public sentiments toward AV advancement in China. The survey's revelations could enrich ethical dialogues, shape regulatory frameworks, and foster increased public involvement in AV technology evolution. Understanding Chinese preferences concerning AVs is crucial for defining ethical norms and legal standards amidst China's rapid technological progress. By addressing gaps in current research, this survey contributes essential data to the global discourse on AV ethics and regulations.

Significance, Limitations and Future Research

As previously noted in this introduction, AV technologies have seen rapid development and extensive road testing in Chinese communities. However, comprehensive ethical and legal research has not kept pace with these advancements. This book project represents the initial significant effort to bridge this gap. While there are ethical and legal principles with universal applicability that can offer broad guidance for the deployment and regulation of AVs in China, not all pertinent issues are culturally neutral. This project delves into Chinese ethical and legal standpoints rooted in Confucian and Daoist values that continue to shape Chinese ethics, influence lifestyles, and impact preferences regarding AV technology development. The book's significance lies in substantiating these culturally relevant Chinese perspectives on AV deployment, blending normative analyses with quantitative studies to present a holistic view of how China should approach ethical and legal challenges in the realm of AVs. It serves as a valuable resource for Chinese policymakers in crafting pertinent policies and guidelines to steer the current development and implementation of AVs in Chinese society. Furthermore, it offers instructive insights and resources for other cultures and societies grappling with similar AV issues in their respective contexts.

While our research presented in this book contributes valuable insights, it is important to acknowledge the inherent limitations that shape our findings and directions for future research. One primary limitation lies in the evolving nature of autonomous driving technology itself. The levels of automation attainable at different stages, the evolving characteristics and capabilities of autonomous vehicles, and their adaptation to varying road conditions introduce a level of uncertainty and change. The discussions within this book are founded on the current technological landscape and projections, which are subject to alterations in the future. As a result, this study necessitates continued exploration, ongoing reflection, and updates that align with technological advancements and shifts. Embracing this dynamic nature of autonomous driving technology is crucial for maintaining relevance and accuracy in our research findings.

Another notable limitation pertains to China's current policies, which exhibit a favorable stance towards automated driving advancements. Despite this encouragement, uncertainties persist regarding the practical implementation timeline and the specific domains in which automated driving technologies can be effectively deployed. Questions linger about whether AVs will predominantly serve logistical purposes, revolutionize passenger transport, or navigate intricate road scenarios. These uncertainties introduce complexity into our research, potentially limiting the depth of our discussions within a particular contextual framework. Our analysis operates within a broader policy landscape, potentially overlooking nuanced policy intricacies. Moving forward, there is a need for more targeted investigations that delve into specific policy nuances and implementation strategies to enhance the depth and specificity of our research insights. By addressing these focused policy issues, future studies can provide a more detailed and comprehensive understanding of the implications and challenges associated with the integration of automated driving technologies within the Chinese regulatory framework.

Finally, a crucial limitation of this study lies in its comparative nature between China and Western contexts. While juxtaposing ethical and legal frameworks between these regions offers valuable insights, there is a risk of oversimplification and a dearth of thorough and nuanced analysis. Given the intricate interplay of diverse cultures, ethical norms, and legal systems, it may not be feasible to comprehensively compare all pertinent aspects, thereby constraining the depth of this comparative study. Future researchers should endeavor to address this limitation by conducting more granular and detailed analyses that consider the unique nuances of each culture and legal system. By delving deeper into specific cultural, ethical, and legal facets, future studies can provide a more nuanced and comprehensive understanding of the similarities and differences between China and the West in the realm of autonomous driving technologies.

References

Awad, E., Dsouza, S., Kim, R., Schulz, J., Henrich, J., Shariff, A., Bonnefon, J. F., & Rahwan, I. (2018). The moral machine experiment. *Nature, 563*, 59–64.

Awad, E., Dsouza, S., Shariff, A., Rahwan, I., & Bonnefon, J.-F. (2020). Universals and variations in moral decisions made in 42 countries by 70,000 participants. *Proceedings of the National Academy of Sciences, 117*(5), 2332–2337.

Edelmann, E., Stümper, S., & Petzoldt, T. (2021). Cross-cultural differences in the acceptance of decisions of automated vehicles. *Applied Ergonomics, 92*(1), 1–10.

European Commission. (2020). Ethics of connected and automated vehicles: Recommendations on road safety, privacy, fairness, explainability and responsibility. Publication Office of the European Union. https://op.europa.eu/en/publication-detail/-/publication/89624e2c-f98c-11ea-b44f-01aa75ed71a1/language-lv.

Geisslinger, M., Poszler, F., & Lienkamp, M. (2023). An ethical trajectory planning algorithm for autonomous vehicles. *Nature Machine Intelligence, 5*, 137–144.

Gogoll, J., & Müller, J. (2017). Autonomous cars: In favor of a mandatory ethics setting. *Science and Engineering Ethic, 23*(3), 681–700.

Lawlor, R. (2022). The ethics of automated vehicles: Why self-driving cars should not swerve in dilemma cases. *Res Publica, 28*, 193–216.

Liu, Y. 刘亚宁 (2024). Wuhan's 'driverless' technology is accelerating (武汉"无人驾驶"加速跑), Economic Observer Website (13 July 2024). https://www.eeo.com.cn/2024/0713/672633.shtml (accessed 9 August 2024).

MIT. (2024). Moral machine. https://www.moralmachine.net/ (accessed 10 August 2024).

Nyholm, S., & Smids, J. (2020). Automated cars meet human drivers: Responsible human-robot coordination and the ethics of mixed traffic. *Ethics and Information Technology, 22*, 335–344.

Porter, Z., Habli, I., McDermid, J., & Kaas, M. (2024). A principles-based ethics assurance argument pattern for AI and autonomous systems. *AI and Ethics, 2024*(4), 593–616.

Rhim, J., Lee, J. H., Chen, M., & Lim, A. (2021). A deeper look at autonomous vehicle ethics: An integrative ethical decision-making framework to explain moral pluralism. *Front Robot AI.* https://doi.org/10.3389/frobt.2021.632394

Tang, J. 唐俊. (2024). Why is the world's largest autonomous driving operation service area in Wuhan? (全球最大自动驾驶运营服务区，为什么在武汉), Sina Finance Website (17 July 2024). https://finance.sina.com.cn/jjxw/2024-07-17/doc-incemyqt7002315.shtml (accessed 8 August 2024).

Wan, Y., Akpan, E., & Guo, H. (2022). The cultural influence of control sharing in autonomous driving. *International Journal of Technoethics, 13*(1), 1–13.

Xiao, M. 肖梦吟 (2024). 'The Wuhan Intelligent Connected Vehicle Development Promotion Regulations (Draft) entered the review stage of the Standing Committee of the Municipal People's Congress' (《武汉市智能网联汽车发展促进条例(草案)》进入市人大常委会审议阶段), Economy and Information Technology Website (19 April 2024). https://www.enicn.com/Enicn/2024/article_0419/75233.html (accessed 8 August 2024).

Xie, H., Liu, Y., & Sun, Y. 谢慧敏、刘逸鹏、孙亚云 (2023). Wuhan accelerating to build itself into the first city for 'autopilot' and 'double breaking 1,000' for the license issuance and opening road (武汉加速打造"自动驾驶"第一城 牌照发放与开放道路"双破千"), Wuhan Government Website (7 September 2023). https://www.hubei.gov.cn/hbfb/szsm/202309/t20230907_4828112.shtml (accessed 9 August 2024).

Xun, Y., & Gong, L. 熊琦、龚联康 (2024). How did Wuhan become the city of automated driving? (武汉这一自动驾驶之城是如何炼成的?), Beijing Daily Website (12 June 2024). https://news.bjd.com.cn/2024/06/13/10803114.shtml (accessed 8 August 2024).

Xun, Z. 熊展平 (2024). Experts suggest Wuhan build 'China's first city for autonomous driving' (专家建议武汉打造"中国自动驾驶第一城"), Wuhan Government Website (22 June 2024). https://www.wuhan.gov.cn/sy/whyw/202406/t20240622_2419319.shtml (accessed 8 August 2024).

Zhou, C., & An, Y. 周春媚 安宇飞 (2024). Competing in a new track, Wuhan strives to become the 'first city for autonomous driving' (逐鹿新赛道 武汉奋力冲击"自动驾驶第一城"), STCN Website (28 June 2024). https://stcn.com/article/detail/1243159.html (accessed 8 August 2024).

PART II

Chinese Ethical Accounts

CHAPTER 2

Automated Vehicles, Human-Machine Relation, and Ethics: A Daoist View

Qian Wang and Wei Zhang

Introduction

The automation of machinery is a dream that human beings have been pursuing all the time. The original meaning of the automobile is "automatic walking device" (auto + mobile), and its invention solved the problem of automation of power. Then humans began to "eliminate the last weak link in the automotive application chain—the driver" (Schwartz & Karen, 2018, p. vii), in order to achieve the automation of driving, and the automated automobile is precisely the outcome of this ideal of complete automation, embodying a dual-layered "automation". The first layer of automation involves the automation of the power system, while the second layer involves the automation of the operating system. As automation has altered the traditional interaction between humans and machinery, it has given rise to a new paradigm of human-machine interaction. Consequently, automation has triggered new contemplations regarding

Q. Wang (✉) • W. Zhang
Department of Philosophy, Dalian University of Technology, Dalian City, China
e-mail: qianwang@dlut.edu.cn; zhwei@dlut.edu.cn

C. Ding, R. Fan (eds.), *Chinese Ethical and Legal Perspectives on Automated Vehicles*,
https://doi.org/10.1007/978-981-96-1277-2_2

the relationship between humans and machines. When Wiener introduced "Cybernetics" years ago, he had a forward-thinking consideration of the "usefulness of humans", as evidenced by his book titled "The Human Use of Human Beings: Cybernetics and Society", which precisely explores how humans and machines collaborate. The rapid progress of contemporary automated vehicles has outpaced the era of Wiener, bringing to the forefront ethical concerns regarding human-machine interactions and garnering significant attention from diverse sectors of society.

To examine these ethical issues through the lens of Daoism and provoke contemplation on technological progress, this article will primarily delve into the societal ethics of contemporary automated vehicles from the perspective of Daoist philosophy of technology, with the goal of offering valuable insights. Firstly, we will contemplate the ethical dilemmas in the human-machine relationship sparked by automated vehicles from the perspective of Daoism. Then, we will explore what "Dao" signifies in automated vehicles based on Daoist technological viewpoints. Finally, we will investigate the approaches and methods to achieve "governing technology by Dao" in the field of automated vehicles from the perspective of Daoist technological philosophy.

The Ethical Dilemmas of Automated Vehicles from the Perspective of Daoism

The Daoist view on the relationship between humans and machines is prominently reflected in a fable in *Zhuangzi*:

> After his travel to the state of Chu in the south, Zigong returned to the state of Jin. When he was passing by the southern bank of the Hanshui River, he saw an old man working in his vegetable garden. From a tunnel that reached the bottom of the well, the old man was fetching water with a jar to water the fields. He had been working hard at it but by no means efficiently. Zigong said, 'There is a kind of machine with which you can water a hundred plots a day. You don't have to work so hard but will work much more efficiently. Don't you want to have one?'
>
> The old gardener looked up and said, 'How does it work?'
>
> Zigong said, 'The machine is made of wood that is heavy in the rear and light in the front. It draws water like pouring, as if the water was boiling over. It is called a well-sweep.'
>
> His face became red with anger, and the gardener grinned and said, 'I've learned from my teacher that a man who employs an ingenious machine

> must do mechanical affairs (*ji shi*, 机事) and that a man who does mechanical affairs must have a mechanical mind (*ji xin*, 机心). A man with a mechanical mind will not be pure and simple; a man without a pure and simple mind will have an unrestful spirit; a man with an unrestful spirit will not attain Tao. I do not use a well-sweep not because I don't know that it works well, but because I feel it a shame to use it.' (Zhuangzi, 1999, p. 187)

Similar to this, Laozi (571 BC–471 BC) also made statements such as "The more devices that bring gain, the more the clan and kingdom's bane. The more that clever crafts expand, the stranger artifacts abound" (Laozi, 2001, p. 145). Many people regard this statement as evidence of Daoist's opposition to machinery application and technological progress. However, does Daoism unconditionally oppose any mechanical application and technological progress? Joseph Needham once pointed out that Daoism had more positive influence than other schools in the history of Chinese science and technology (Needham, 1991, p. 41). How to explain the relationship between this historical phenomenon and the Daoist view of technology? Here it involves how to understand the specific meaning of "mechanical affairs" and "mechanical mind" of Daoism.

People in Zhuangzi's time using the well-sweep and modern people using machinery (including automated vehicles) both have the same effects essentially, that is, time-saving, labor-saving, and outstanding benefits. It is certainly necessary to relieve the heavy physical and mental labor. However, if the use of machinery (including a variety of tools and machines) has a negative impact on people's physical and mental health, then this kind of mechanical application should be restricted. Safeguarding the physical and mental health of individuals necessitates regular exercise, a suitable natural environment, and a harmonious social environment. It was through labor that humans evolved from apes, with labor playing a pivotal role in shaping human physiological functions and spiritual vitality. The advent and development of machinery have been intertwined with the necessity for labor, albeit not always in perfect harmony. Machinery can assist or replace human labor, but the use of machinery has its own laws and external conditions, which do not fully conform to the physiological and psychological characteristics of users. Because people use machinery mainly for production and economic considerations and can obtain significant benefits with as little investment as possible (this is the capital attribute of machinery), the demand for machinery in the entire society does not only depend on the individual choices of the user (or consumers) but are constrained by a variety of social factors.

Generally, the owners of tools or machinery (such as well-sweep, water mills, carriage, etc.) were also operators and beneficiaries in the agriculture and handicrafts era. In the modern factory system, tools and machinery have transformed into diverse machines, with factory owners emerging as the primary beneficiaries, while operators are merely hired workers who receive a small portion of the benefits. As for other stakeholders involved in machine applications, such as consumers, managers, the public, etc., there are various complex relationships between them and the economic benefits of machine use. The adoption of machines has become a pivotal element in shaping modern society. People's clothing, food, housing, and transportation are all at the mercy of machines, and machines have become what Heidegger called the Gestell (Heidegger, 1997, pp. 3–35). Many people may not like the use of certain machines, but they have to obey the arrangements of society, just as Chaplin showed workers being at the mercy of machines in the movie "Modern Times". And many elderly people may face the predicament of not being able to use the "health QR code" (健康码) with mobile phones during the COVID-19 prevention period. Some hospitals overtreat patients, and use various instruments and equipment to detect physiological indicators for patients (Du, 2005). In fact, some of them are unnecessary, and they are often just designed to benefit doctors and hospitals.

Unreasonable application of machinery is likely to cause various negative effects. If the deployment of machinery contradicts people's physical and mental health, as well as the social and natural environment, it is a "mechanical affair". If people's health, life, communication, and the development of the whole society are completely subject to the regulations and guidance of "mechanical affairs", then the "mechanical mind" is at work. At this time, "humans" are enslaved by "things", which is abnormal. As Roland Barthes said," I think that cars today are almost the exact equivalent of the great Gothic cathedrals: I mean the supreme creation of an era, conceived with passion by unknown artists, and consumed in image if not in usage by a whole population which appropriates them as a purely magical object" (Barthes, 1991, p. 88). In this case, the warnings to "mechanical affairs" and "mechanical mind" in the fables of *Zhuangzi* are not unreasonable.

How to distinguish "mechanical affairs" and "mechanical mind" from people's normal use of machinery? Going back to the era when Zhuangzi

(369 BC–286 BC) lived, did it make sense that the old man regarded the use of the well-sweep as "mechanical affairs" and "mechanical mind"? If we limit our consideration to the relationship between the old man, the well-sweep, the urn, and the vegetable field, and only consider the functions and benefits of different tools and usage methods in watering, it is easy to conclude that Daoists reject all mechanical applications as "mechanical affairs" and "mechanical mind". However, if we expand our perspective and take into account additional factors, then we have to pose the question: would it be beneficial for the elderly if the well-sweep is not used? If the elderly individual is in good health, he can not only hug the urn to irrigate the field but also enjoy himself, which is good for his physical and mental health, why does he have to use the well-sweep? Of course, he will feel that the use of the well-sweep is closely related to "mechanical affairs" and "mechanical mind", so he is "ashamed to do it". The criticism of "mechanical affairs" and "mechanical mind" in the fable of *Zhuangzi* should be aimed at this situation. If we switch the situation to the present day and think a little more extremely, we suppose that Zigong no longer recommends the old man use the well-sweep, but instead use the contemporary automatic watering system, which will automatically water the ground at the touch of a switch (of course it can also realize automatic farming, automatic seeding, automatic fertilization, automatic weeding, automatic harvesting), the old man can completely sit back and enjoy the fruits. Is this the best choice for him? Indeed, some people's dream is to "reap where one has not sown", but for most people, "reap where one has not sown" is a negative evaluation. If a modern farmer fully embraces an automated watering system and mechanizes all farming processes, with his sole task being to press a switch, can he still be considered a "farmer" in the original meaning of this word? Is this mode of production the "happiness" that he should pursue? Furthermore, how does he safeguard his own health? Does he need to go to the gym? Would it be better for his health if he farmed? We can find out kinds of benefits of proper farming on people's physical and mental health, such as fresh air, enhanced metabolism, proximity to nature, joyful mood, sense of accomplishment, etc., and these benefits cannot be replicated by automated machinery.

Of course, if the old man is not in good health and can't hold the urn to water the field, it is indeed pedantry not to use the well-sweep. The fable in *Zhuangzi* pushes the criticism of "mechanical affairs" and "mechanical mind" to the extreme, and the purpose is to make people vigilant, but the implicit premise is that not using the well-sweep does not actually cause pressure on the old man. The Daoist view of technology does not exclude the normal use of tools and machinery in an absolute sense, otherwise, there would be no such allusions as Carpenter Shi Swing axe (*jiang shi yun jin*, 匠石运斤), Ziqing Making Ju (*zi qing zuo ju*, 梓庆做鐻), and Lunbian Chopping Wheel (*lun bian zhuo lun*, 轮扁斫轮). Of course, in these allusions to the Daoist concept of technology, the relationship between man, tool and machine is closely coordinated or even integrated, and the criticism and warning of "mechanical affairs" and "mechanical mind" are not contradictory to this.

Let's imagine such a thought experiment. If Zhuangzi lived in our time, what kind of comments would he make when he sees an automated vehicle? Automated vehicles and the well-sweep (*jie gao*, 桔槔) in Zhuangzi's fable are both mechanisms, so Zhuangzi's discussion of automated vehicles can also be regarded as a modern version of the fable cited above. When Zhuangzi sees the advantages of automated vehicles demonstrated by modern people, he may ask: "Why do humans have to use automated vehicles?" Automated vehicles undoubtedly fulfill practical requirements, offering convenience to users and proving profitable for production and sales companies. However, the application of automated vehicles will lead to a series of social, ethical, and legal issues, such as whether the development of automated vehicles is to allow people to enjoy the convenience of travel, or to be a monopolistic means of transportation. Will the automated system become a lethal weapon once it is hacked (Schwartz & Karen, 2018, pp. v–vii)? Can the safety configurations promised by the manufacturers be fulfilled, and will things like the "emission gate"[1] happen again? When faced with the "trolley dilemma", should the lives of the few be sacrificed to save the lives of the many? How to coordinate the conflict of interests between new industries and traditional industries? How to resolve the conflict between convenience and security? What ethical principles should the algorithm of autonomous driving follow? Who will be responsible if automated vehicles crash? How to change the current traffic regulations and property rights system to meet the challenge of autonomous driving? etc. If we don't take these

[1] On September 18, 2015, the U.S. Environmental Protection Agency accused Volkswagen of selling some diesel vehicles with software specially designed to deal with exhaust emission detection, which can identify whether the vehicle is under detection, and then start it secretly during the vehicle inspection, thereby it enables cars to pass muster with "high environmental protection standards" during vehicle inspections, but when driving in normal times, these cars emit a large amount of pollutants, up to 40 times the legal standard of the United States.

factors into account and just think about making money with automated vehicles, that's "mechanical affairs" and "mechanical mind".

The fable mentioned above and the ethical dilemma of automated vehicles both touch upon themes of "machinery", "mechanical affairs" and "mechanical mind", and ultimately "will not attain Dao". To grasp the Daoist perspective on technology, it is insufficient to solely examine the relationship between humans and machines or to label the old man as a symbol of resistance to technological advancement. Similarly, the ethical considerations of automated vehicles should not only focus on the dynamics between passengers and pedestrians but also encompass the interactions among designers, manufacturers, marketers, managers, scholars, and the general public. Viewing the ethical dilemma of automated vehicles from the perspective of Daoist technology, we must consider all these stakeholders, even including what Latour called "non-human" actors (Latour, 2006), and this leads to how to understand the "Dao" of automated vehicles.

The "Dao" of Automated Vehicles from the Perspective of Daoism

"Dao" is the core category of Daoism, and its meaning is extraordinarily rich and profound, which can be interpreted from multiple angles and levels. Then, from the perspective of the philosophy of technology, how should "Dao" be understood? Let's look at another fable from Zhuangzi that people are familiar with:

> A butcher was carving a bullock for Lord *Wenhui*. At every touch of his hand, every move of his shoulder, every stamp of his foot and every nudge of his knee, there came the sound of slicing the flesh and wielding the knife—a perfect rhythm to the Dance of Mulberry Trees and a perfect tune of the music in King Yao's time.
>
> Lord *Wenhui* remarked, 'Qh, splendid! That you have such a masterful skill?'
>
> The butcher put down his knife and responded, 'What I lyre is Tao, which is much more splendid than my skill. When I first began to carve a bullock, I saw nothing but the whole bullock. Three years later, I no longer saw the bullock as a whole but in parts. Now I work on it by intuition and do not look at it with my eyes. My visual organs stop functioning while my intuition goes its own way. In accordance with the natural grain, I cleave along the main seams and thrust the knife into the big cavities. Following

the natural structure of the bullock, I never touch veins or tendons, much less the big bones? A good butcher Changes his knife once a year because he cuts the flesh; an ordinary butcher changes his knife once a month because he hacks the bones. Using this knife for nineteen years, I have carved thousands of bullocks, but the edge of my knife is still as sharp as if it had just come from the whetstone. There are crevices between the joints, but the edge of my knife is very thin. When I insert the thin edge of my knife into these crevices, there is plenty of room for it to pass through, that is why, after nineteen years, the edge of my knife is still as sharp as if it had just come from the whetstone. Nevertheless, whenever I come to a complicated spat and see that there are difficulties here, I proceed with great caution. 1 keep my eyes on what I am doing and wield the knife slowly. With a very slight movement of the knife, I cut off the flesh—it falls to the ground like a clod of earth. Holding the knife in my hand, I stand there, leaking proudly around. Then I clean the knife and put it carefully away.'

Lord *Wenhui* said, 'Splendid! From what you just said, I've learned how to keep good health'. (Zhuangzi, 1999, pp. 43–45)

The original meaning of Dao in Chinese is "road", which implies the act of walking on a specific path guided by the mind. Later, its meaning was extended to include the operation steps, procedures, and methods in various practical activities. The fable of "Butcher Ding Carving an Ox"(*Pao Ding Jie Niu*, 庖丁解牛) reveals that Daoism abstracts this essential feature in human's practical activities, and proposes the category of "Dao", intending to show that any practical activity has its most reasonable and optimal operating procedures, the concept of "Dao" in Daoist philosophy of technology is actually an ideal state in technical operation activities, where there is a pursuit of perfect alignment in nature, structure, and functionality among the operator, the object of operation, and the tools involved, entirely avoiding any contravention of natural laws through artificial intervention. Laozi's "The Great Dao is formless" (*da dao wu xing*, 大道无形) refers to this idealized state. When the unity of nature and man (*tian ren he yi*, 天人合一) and the integration of things and self (*wu wo jiao rong*, 物我交融) are achieved, there is no trace of deliberate action. Especially, the action of Pao Ding should consider the physiological characteristics of the cattle and the physical characteristics of the knife, as well as the physiological and psychological characteristics of himself. There is a process of enhancing his skills, from the incomplete adaptation between the operator, the object of operation and the tool to the ease of use and perfection, so that his physical and psychological potential can be fully exerted, and the operation of unblocking cattle can be raised to the level of an almost artistic performance. Pao Ding is essentially able to obtain the most significant benefits (fast and good, physical, and mental pleasure, and a sense of accomplishment) with the smallest investment (saving time, effort, and knife). To this end, it is necessary to adopt a series of the most reasonable and optimized operation procedures (cut the knife at the place with ease, relying on in-depth experience to grasp the sense of propriety and pursue just right).

The fable of "Butcher Ding Carving an Ox" helps us to understand the meaning of "self-momentum" (*zi ran*, 自然)[2] and "non-action" (*wu wei*, 无为) in Daoism as a whole. If we think about them within the framework of modern Western philosophy, *Zi Ran* in Daoism can easily be misinterpreted as (1) a synonym for "nature"; (2) "the primitive social state in human history"; (3) "a state out of human civilization"; (4) "a state devoid of any human effort"; (5) a "state of nature" in the Hobbesian sense. In fact, *Zi ran* of Daoism emphasizes "naturalness", that is, things present a harmonious state of internal and external relations according to their own characteristics. Natural phenomena such as volcanoes, earthquakes, and tsunamis are full of conflicts and confrontations. Laozi's "Dao follows *Zi ran*" obviously does not follow such "nature". Liu Xiaogan thinks that the *Zi ran* mentioned by Daoists is a "humanistic nature" (*ren wen zi ran*, 人文自然), which is reasonable, because it embodies the pursuit and yearning of human beings. *Zi ran* mentioned by Daoists "is not the nature of heaven and earth, physical nature, biological nature, barbaric state, primitive stage, and anti-cultural and anti-civilization concepts" (Liu, 2006, pp. 48–49). It "expresses the ideal and pursuit of the living conditions inside and outside the human group, and the yearning for the harmony and order of nature" (Liu, 2006, pp. 48–49). However, emphasizing *Zi ran* doesn't mean denying and rejecting any influence of external factors. Because there is no individual absolutely free from the influence of external factors, as long as the action of external forces does not arouse people's direct or strong feelings, it can be counted as nature (Liu, 2006, p. 210). So Pao Ding's carving of the bullock is such a process of conforming to *Zi ran*.

[2] In English, *Zi ran* is translated to "nature", "naturally so", or "self-So", etc. In this paper, we choose "self-momentum" (translated by Moss Roberts), because it is a better translation to express the meaning: The way things are, or all as it is given.

If the concept of *Zi ran* is understood in this way, then *Wu wei* will not be misunderstood as "doing nothing at all" (*quan wu shi shi*, 全无事事) (Li, 1986, p. 537). In the past, it was often believed that "non-action" (*wu wei,* 无为)[3]referred to the general actions, behaviors, activities, and motions, seemingly encompassing all human activities. Therefore, "non-action appeared to negate all human activities (Liu, 2006, p. 388). As a result, discussions about "mechanical affairs" and "mechanical mind" in Daoism were often interpreted negatively. People mistakenly assumed that Daoism universally rejects technology, which made Daoism appear excessively conservative and outdated. This is a significant misunderstanding of Daoist philosophy. As mentioned before, *wu wei* seems to be a universal judgment, as if it negates all actions, but in fact, *wu wei* only negates "the usual and conventional behaviors, especially those that cause conflicts, must pay a huge price and have a bad effect" (Liu, 2006, p. 388). This point can be confirmed from many expressions in *Dao De Jing*, such as "Live through their lives without his possession; And act of themselves, unbeholden to him; To the work he completes, he lays down no claim" (生而不有，为而弗恃，功成而弗居), "With the fruits of victory desist, Never seek to break a beaten foe" (善有果而已，不敢以取强), "A boon to every living creature, In adverse relation never" (水善利万物而不争), "The wise know how to salvage men, and make sure none go to waste; They know how to salvage things, and make sure none go to waste" (是以圣人恒善救人，故无弃人；恒善救物，故无弃物), "Virtue well-founded nothing uproots, Virtue fast-bound no one can steal" (善建者不拔，善抱着不脱), "Expert marching leaves no trails; Expert wording has no flaws" (善行无辙迹，善言无瑕谪), "In this way support and sustain the self-becoming of the myriad, and do not presume to act upon them" (以辅万物之自然,而不敢为), in which "live" (*sheng*, 生), "work" (*gong*, 功), "fruit" (*guo*, 果), "boon" (*li*, 利), "salvage" (*jiu*,救), "found" (*jian*, 建), "marching" (*xing*, 行), "wording" (*yan*, 言) and "support" (*fu*, 辅) are not "action" (*wei*, 为)? These are all "actions" that are recognized by Daoism, these actions are actions that leave no traces, such as the "action" that Pao Ding's carving of the bullock hardly wears out knives. The key point of "non-action" is not to control and manipulate things in a coercive way, but to let all things themselves according to their own nature. In a word, the "action" that Daoist opposes is the action of "forcing".

[3] *Wuwei* also can be translated into different English words according to its contexts, such as "minimal action", "under-govern", "without leading", "not striving", or "pursue no end".

It can be seen from this that "self-momentum" and "non-action" are two sides of the "relationship between heaven and man" (天人关系). "Self-momentum" focuses on the aspect of "heaven", while "non-action" focuses on the aspect of "human". Therefore, Daoism certainly doesn't unconditionally oppose all mechanical applications and technological progress but only opposes unnatural technology, because unnatural technology is bound to be accompanied by rebound, resistance, and destructive conflict, so it is unsustainable in the long run. Conversely, if manpower is used to assist the natural world and follow the internal motivation of nature, it is a natural technology. Pao Ding's "accordance with the natural grain", unlike *Liang Pao* and *Zu Pao's* "cutting" and "folding", reflects the original meaning of "self-momentum", that is, conforming to the nature of things and no force. Pao Ding's operation has become "self-momentum" in the Daoist sense, which seems to be "action", but there is no trace, so it is the ideal state of human activities including technical activities.

The Daoist emphasis on "self-momentum" and "non-action," and opposition to "forcefulness" and "reckless action" reflects a pursuit of the "constant" (*chang,* 常) and the "long-lasting" (*jiu,* 久). *Dao De Jing* contains numerous expressions regarding the "constant" and the "long-lasting," such as "The Dao that can be spoken is not the eternal Dao; The name that can be named is not the eternal name," (道可道，非常道；名可名，非常名) "Going back to one's Destiny is to find the Eternal Law. To know the Eternal Law is Enlightenment. and not to know the Eternal Law is to court disaster. He who knows the Eternal Law is tolerant; Being tolerant, he is impartial; Being impartial, he is kingly; Being kingly, he is in accord with Nature; Being in accord with Nature, he is in accord with Tao; Being in accord with Tao, he is eternal, and his whole life is preserved from harm." (复命曰常，知常曰明。不知常，妄作凶。 知常容，容乃公，公乃全，全乃天，天乃道，道乃久，没身不殆) "The Dao's constant virtue is to be sufficient; therefore, true return leads to the great integrity," (常德乃足，复归于朴) "he who is contented with contentment shall be always content." (故知足之足，常足矣。) "This is called following the constant," (是为习常) "Knowing harmony is called the constant; knowing the constant is called clarity," (知和曰常，知常曰明), " he who does not lose his center endures" (不失其所者久) "who know when to stop runs into no danger, He can long endure." (知止不殆，可以长久) "the mother (principle) of a ruling country can long endure. this is to be firmly rooted, to have deep strength, the road to immortality and enduring vision" (有国之母，可以长久；是谓深根固柢，长生久视之道) (Lin, 1948).

These expressions highlight Daoism's profound focus on the concepts of "constant" and the "long-lasting." It could be argued that the pursuit of the "constant" and the "long-lasting" is the ultimate goal of Daoism, with other aspects serving as means to achieve this end. "Self-momentum" and "non-action" align with the intrinsic properties and laws of things, refraining from undue interference in things, thereby enabling their stable and continuous existence. Conversely, not adhering to the inherent rules of things and interfering with their development might yield certain short-term effects but, fundamentally opposing the natural laws, will ultimately face repercussions, returning things to their original state. Consequently, the results of "forcefulness" and "reckless action" are always transient and unpredictable.

Similarly, concepts like "accomplish tasks and withdraw," (功遂身退) and "knowing white, keep to black," (知白守黑) which seem like passive and conservative behaviors, are actually geared towards the goal of the "constant" and the "long-lasting." According to the dialectical logic of "reversion is the action of Dao," (反者道之动) when anything reaches its peak, it will inevitably head towards its opposite. When a person achieves success, if he/she cannot promptly adjust his/her mindset, he/she might become complacent in his/her past successes and stagnate, thereby heading downhill and failing to maintain his/her success, thus unable to sustain it in the "constant" and the "long-lasting." Recognizing this, Daoism advocates "accomplish tasks and withdraw," and "knowing white, keep to black," positioning oneself in a "weak" state, consistently acknowledging one's shortcomings, leaving room for future development, refraining from self-conceit, and thus maintaining space for continual progress. This way, one can transition from one success to another, making success "constant" and "long-lasting" rather than a mere fleeting moment.

To attain the objective of "self-momentum" and "non-action", it is essential to refrain from arbitrary actions that go against natural laws. Ancient Chinese philosophers developed the concept of "governing technology by Dao", which means guiding the development of "techniques" according to the requirements of the "Dao" (Wang & Zhang, 2019, pp. 133–148). The idea of "governing technology by Dao" is most apparent in Daoism. However, among the pre-Qin philosophers, Confucianism, Mohism, Legalism, and others also contained the concept of "governing technology by Dao". While the core of this concept closely resonates with

Daoist understanding, there are variations in emphasis. For instance, the pre-Qin Confucians' notion of "governing technology by Dao" directly addressed the societal effects of technological application, emphasizing the macro-level societal effects of engineering technology, striving to limit and eliminate the negative impacts brought about by inappropriate technological applications. In technological endeavors, Confucianism emphasized not only the coordination among operators, objects of operation, and tools, but also the interconnectedness of technological activities with elements such as nature, society, and ethics. The Mohists' expression of "governing technology by Dao" in technological thought emphasized constraining the technical activities of groups or individual craftsmen with ethical and technical standards. It demanded disciples to learn the spirit of Da Yu's hard work in water conservation, without ambition for fame or gain, dedicating themselves selflessly and enduring hardship for the sake of urgent needs in the world. The Legalist thoughts on laws and regulations had a certain normative role in technological development. One of the fundamental meanings of "law" was technical standards. As stated in *Guanzi*: "Measurements, plumb lines, rules, scales, measures, and angles, are called laws" (Liu & Jia, 2016, p. 47). According to Joseph Needham, the rise of the Qin state was closely related to certain military technological inventions. Moreover, the emergence of these inventions cannot be disregarded without considering the background of emphasizing laws and regulations. This is also part of "governing technology by Dao", emphasizing stringent requirements for technical standards.

Some scholars equate "Dao" with the Western philosophical concept of "Logos" and argue that "Dao" is the primal origin of the world, an objective law, and an ineffable and mysterious entity. This overlooks the essential connection between "Dao" and human practical activities. In fact, the relationship between "Dao" and the concept of "program" in modern information science is more closely related. However, "program" is a neutral concept with distinctions between good and bad, whereas "Dao" represents an ideal state of a "program." From this perspective, we can re-understand what Zhuangzi meant by the "mechanical mind" and why it was said that possessing the "mechanical mind" leads to a lack of readiness, an indeterminate spirituality, and something that cannot be conveyed by the Dao. "Mechanical affairs" refer to the mechanical application after losing the constraint of the "Dao," while the "mechanical mind" serves as the ideological foundation of these "mechanical affairs."

According to this understanding of the "Dao" of technology, the "Dao" of automated vehicles should encompass the comprehensive coordination of various relevant elements (nature, society, people, tools, objects) in its technological development. It represents the most rational, optimized approach, method, and process adopted within its technological practice. Only in this way can we achieve the sustainable development of "constant" and "long-lasting" in the field of automated vehicles. However, the current level of automated vehicles cannot guarantee sufficient harmony among users, stakeholders, tools, objects, and environmental elements, but contains many conflicts. Is it reasonable to design, manufacture, and use automated vehicles for utilitarian purposes, and to drive on the road without adequate security? Some people may also argue that automated vehicles are a new thing and should be allowed for a trial, but safety must be the top priority before any trials. Safety concerns cannot be resolved through mere experimentation. In addition, once an accident occurs, it is still difficult to clearly delineate responsibilities, and the decision-making system of the automated vehicles is also difficult to deal with various emergencies in the daily environment. Pedestrians on the road may find it challenging to prevent accidents caused by malfunctioning automated vehicles. If automated vehicles do not show irreplaceable advantages in protecting the environment and promoting sustainable social development, but instead bring uncertain risks, then preventive governance should be strengthened when developing automated vehicles. Now, it is essential to examine whether the Daoist philosophical perspective on technology holds validity and what insights and inspirations it offers for the healthy development of contemporary automated vehicles.

How to Achieve "Governing Technology by Dao" in the Field of Automated Vehicles

From the perspective of the philosophy of technology, understanding the "Dao" as the "most reasonable and optimized" program is akin to the traditional philosophical interpretation of "Dao" as the essence of "self-momentum". Isn't "self-momentum "the highest pursuit of Daoism, representing the most rational and optimized state? In other words, "most reasonable and optimized" are further interpretations of the essence of "self-momentum". They differ only in their expressions, both fundamentally aiming for "harmony." Therefore, "Dao" is a normative category; its guidance and regulation of technical activities embody the concept of "govern technology by Dao". Technical activities, akin to "Pao Ding's carving of the bullock", continually strive to approach this ideal state infinitely. So, how do automated vehicles achieve the principle of "govern technology by Dao"? It should gradually expand the field of vision to all stakeholders, and establish explicit and implicit organic connections between various related elements in discussions of the "Dao" of automated vehicles. The state that conforms to the "Dao" in the field of automated vehicles should truly allows users to feel completely at ease. Here, the feelings and safety of passengers and pedestrians are fully considered, and there is no "calculation" (*suan ji*, 算计) that violates the legitimate rights and interests of stakeholders. There is no buck-passing in the event of an accident, and there is no management omission and coercion. Automated vehicles should not be the predominant or sole mode of transportation, but rather demonstrate their distinctive advantages and appeal in meeting the diverse needs of individuals. To accomplish these objectives, they should adhere to the following three fundamental principles.

The first principle is "Human-Centeredness." (*yi ren wei ben*, 以人为本) Automated vehicles should prioritize essential human interests, safeguarding physical and mental well-being, human dignity, holistic development, and the harmonious interaction between technological advancements and society, as well as nature. If automated vehicles impede or harm human nature, violate human dignity, or have adverse effects on society and the environment, such innovations should be prohibited. Just as drugs, performance-enhancing substances in sports, biochemical weapons, and others, no matter how high their technological level, they cannot be allowed to proliferate. These instances should be regarded as deviations from the "Dao" of high technology and clearly not the original intention of human development in advanced technology. Many pre-Qin era philosophers focused on the issue of "human-centeredness." Daoist theories emphasize that technological activities should harmonize with nature, preserving and nurturing human "innate nature." When using tools or machinery as mere "objects," individuals should not be overwhelmed by them or enslaved by them. In essence, we must reject the misconception that every high-tech advancement is inherently beneficial to humanity.

The second principle is "Perceiving with the Heart" (*yi xin guan wu*, 以心观物). The term "heart" is a profound concept akin to the "Dao." According to traditional Chinese philosophy, the "Dao" needs to be understood and grasped through the "heart." Here, "heart" does not refer to the anatomical heart in the modern sense, but rather the cognitive organ through which humans engage in cognition, psychology, moral judgments, and aesthetic activities, just as ancient Chinese philosopher Wang Yangming said, "The heart is not a piece of flesh; wherever there is perception, there is the heart" (Wang, 2008, p. 390). Traditional Chinese culture stresses the importance of comprehending things from a holistic, organically connected, experiential perspective. Due to the strain, this mental activity can put on the heart when overburdened, ancient Chinese scholars advocated that "the function of the heart lies in thinking," defining the "heart" as the sum of the human body's organs with cognitive functions (Yang, 2014, p. 208). Western culture has traditionally placed a strong emphasis on logical reasoning and quantitative analysis, with highly developed instrumental rationality, but often relegating holistic, organic, experiential thinking to a secondary position. The understanding of the "heart" in traditional Chinese culture helps to compensate for this weakness. The development of automated vehicles increasingly replaces human logical reasoning and quantitative analysis, yet the ability to think by "heart" remains irreplaceable by automated vehicles. In other words, the thinking capability involving the "heart", unique to humans, cannot be replicated by automated vehicles. Therefore, the relationship between human intelligence and automated vehicles naturally translates into a relationship between the "heart" (*xin*, 心) and "objects" (*wu*, 物). Daoism emphasizes that the "heart" should play a leading role in relation to "objects," and "objects" should not dominate the "heart." Otherwise, it would become a "mechanical mind," where utilitarian material demands overshadow the purity of the "heart."

The perspective of "Perceiving with the Heart" is indeed different from specialized, localized, and singular logic analysis and technological applications. It helps prevent technological inventions and applications from becoming what the German philosopher Heidegger referred to as a "gestell" that manipulates and controls people (Heidegger, 1997, pp. 3–35). The "heart" will sensitively perceive these alienating phenomena on a holistic level, actively eliminating their negative influences. This is a form of wisdom that can control automated vehicles without being dictated by it. By employing the "heart" to understand the implementation of automated vehicles, one can discern which applications are beneficial to humanity and which may lead to harmful consequences, enabling people to maintain a vigilant and cautious attitude towards the use of automated vehicles. Besides mitigating negative impacts, "Perceiving with the Heart" will also bring about a wealth of creative positive outcomes, as holistic, organic, experiential cognitive approaches are sources of wisdom and inspiration. In the era of artificial intelligence, "Perceiving with the Heart" requires the organic integration of holistic thinking with analytical thinking, uniting intuition with logic to harness the full functionality of the "heart" at a higher level.

The third principle is "Ethics and Benefit Alignment." (*de fu yi zhi*, 德福一致) In the development of automated vehicles, the pursuit of "benefit" doesn't necessarily exclude aspirations for "ethics." However, the pursuit of "benefit" should not transgress the bottom line of "ethics." Otherwise, it would lose the constraints of the "Dao", leading to negative consequences driven by short-term gains. The development of automated vehicles should be guided by a value system and ethical principles based on the comprehensive development of humans. It should respect individual differences, rights, and obligations of humans, ensuring that the development of automated vehicles maximizes benefits while operating within human-controlled boundaries. Nevertheless, during the design, manufacturing, and usage phases, automated vehicles might embed specific value factors, becoming a means for certain individuals or entities to seek undue benefits. These methods are hidden behind the highly specialized, procedural, and standardized facade of automated vehicles, often without users being aware of the infringements. From the perspective of "Ethics and Benefit Alignment," to truly harness the beneficial role of automated vehicles for humanity, it is paramount for designers, manufacturers, and users of automated vehicles to prioritize self-cultivation through ethical principles, thereby ensuring that automated vehicles are not utilized in ways that jeopardize the sustainable development of human society.

When assessing the societal impact of automated vehicles, technical professionals, business managers, and government departments tend to prioritize the aspect of "benefit," whereas humanists and the general public often emphasize the aspect of "ethics." This imbalance might result in an overemphasis on "benefit," overshadowing ethical issues, while an emphasis on "ethics" might miss the core problems inherent in the application of automated vehicles, rendering it ineffective. Hence, ethical assessments of automated vehicles require mutual understanding, active involvement, and close cooperation between technical professionals and ethicists to foster productive dialogues. The contemporary international rise and widespread promotion of the "Responsible Research and Innovation" model provide crucial insights into implementing the requirement of "Ethics and Benefit Alignment" in the development of automated vehicles. The ultimate application of the "Dao" of automated vehicles must be integrated into technological innovation activities. It must promote the healthy development of automated vehicles under the premise of "responsibility," ensuring that various risks and hazards brought about by automated vehicles are promptly resolved, truly making them a powerful tool for the benefit of humanity.

Following these principles, an important aspect to be considered in the development of automated vehicles is that automated vehicles should become one of the choices for people to travel in the future, but not the only choice, we need to leave room for other travel modes, such as walking, cycling, and public transport and so on. If other modes of transportation are forcibly eliminated in the name of developing automated vehicles, then this is not the "self-momentum" of Daoism, but "action" (*ren wei*, 人为) or "force", which violates the principle of "harmony".

The demands of humans are multilayered and multifaceted. Convenience in travel is just one of the human needs. Furthermore, individuals also require regular exercise to maintain their health. Otherwise, the situation depicted in "Wall-E" may appear: Modern Homo sapiens are cared for in every possible way in the intelligent transportation system, and they are so lazy that they don't even want to move a finger. As a result, they all become fat and abnormally bloated (Schwartz & Karen, 2018, p. xii). Secondly, the situations of people are also different. Some individuals may be in poor physical condition and require more comfortable and convenient intelligent transportation, while others who are healthy can utilize travel to increase opportunities for exercise and achieve their exercise goals. As a result, this group of individuals does not have a pressing need for automated vehicles but rather finds walking or cycling more suitable for physical and mental exercise. If manufacturers and marketers forcefully eliminate other modes of transportation, the needs of these individuals will not be met. Furthermore, without fundamentally solving the risk of autonomous driving, passengers and pedestrians will be worried when traveling. At this time, it would be unwise to promote automated vehicle technology without implementing appropriate regulations. Otherwise, people may feel the taste of "mechanical affairs" and "mechanical minds" in the propaganda of automated vehicles. If it is only said that automated vehicles are the product of technological innovation, and innovation represents social progress, it is backward and pedantic to reject the use of automated vehicles. Doesn't this kind of reasoning make people feel that there is an obvious departure from social ethics?

Therefore, the future transportation system should leave optional space for people to choose, "making automatic drive technology become a booster for the improvement of the transportation system, instead of mistakenly taking the old road a century ago—giving up the road entirely to cars" (Schwartz & Karen, 2018, p. 28). In fact, if the automated vehicles are properly used, it can provide an opportunity for the realization of the goal of health, because the control of the driving trajectory of the automated vehicles is very precise, and there is no need for the city streets to be equipped with 3.5-meter-wide lanes for the 1.65-meter-wide cars. The space saved can be used to widen or increase sidewalks, bicycle lanes, and green belts, thereby increasing opportunities for people to exercise. In addition, automated vehicles will also promote the development of the "sharing model", which will increase the utilization rate of cars and reduce the number of private cars, or even private cars will disappear forever. This will greatly reduce the area of parking lots, and the land saved can also be used to increase public activity areas, such as parks, green spaces, entertainment venues, etc.

In essence, if we fully exploit the advantages of automated vehicles and align with the current trend, we can facilitate the harmonious development of automated vehicles within the broader transportation ecosystem. This integration would lead to a reduction in "action" and "conflict," enabling the realization of "self-momentum" and "non-action." Ultimately, this would result in the most "reasonable and optimal" and "harmonious" state between humans and nature, society, and ultimately, the "governing technology by Dao" in the field of transportation systems.

References

Barthes, R. (1991). *Mythologies*. Farrar, Straus & Giroux, The Noonday Press.

Du, Z. 杜治政 (2005). Overtreatment, appropriate medical care and optimization of medical care (过度医疗、适度医疗与诊疗最优化). *Medicine and Philosophy, 26*(7), 1–4.

Heidegger, M. (1997). *The question concerning technology and other essays*. Garland Publishing.

Laozi. (2001). *Dao De Jing: The book of the way*. University of California Press.

Latour, B. (2006). *Reassembling the social: An introduction to Actor-Network-Theory*. Oxford University Press.

Li, J.黎靖德 (1986). *Zhuzi Yulei* (朱子语类). Zhong Hua Book Company.
Lin, Y. 林语堂 (1948). *The wisdom of Laotse*. The Modern Library.
Liu, X. 刘笑敢 (2006). *Laozi from ancient to the modern: Comparative studies of the five versions, including introductory analyses and criticisms* (老子古今:五种对勘与析评引论). China Social Sciences Press.
Liu, X. 刘向, & Jia, T. 贾太宏 (2016). *Guanzi Tongshi* (管子通释). Xiyuan Publishing House.
Needham, J. (1991). *Science and civilisation in China, Volume 2: History of scientific thought*. Cambridge University Press.
Schwartz, S. I. & Karen K.(2018). *No one at the wheel: Driverless cars and the road of the future*. Public Affairs.
Wang, Y. 王阳明 (2008). *Chuan xi Lu* (传习录). Zhongzhou Ancient Books Publishing House.
Wang, Q. 王前, & Zhang, W. 张卫 (2019). Reflection on the Dao of technology: Philosophy of technology from the perspective of Chinese culture. In Q. Wang (Ed.), *Chinese philosophy of technology: Classical readings and contemporary work*. Springer.
Yang, B. 杨伯峻 (2014). *Mengzi Yizhu* (孟子译注). Zhonghua Book Company.
Zhuangzi. (1999). *Zhuangzi* (庄子). Hunan People's Publishing House & Foreign Languages Press.

CHAPTER 3

Algorithmic Morality and Moral Decision Making for Automated Vehicles: A Confucian Account

Jue Wang

Introduction: Algorithmic Morality in the Context of Automated Vehicles

Many people believe that automated vehicles could be superior drivers to humans because they consistently make rational decisions. Although being a rational driver does not necessarily imply being a trustworthy driver, it is important to note that these two attributes are not synonymous. Researchers question whether a machine, no matter how advanced it could be, can make a moral decision in crash cases that involve personal harm (Morning & Meschtscherjakov, 2019). Additionally, individuals exhibit a natural and emotional aversion to allowing machines to make decisions regarding life and death in situations akin to those presented in the trolley problem. Against this background arises "a need for algorithmic morality, broadly construed as a set of variables and action scripts, that could decide

J. Wang (✉)
Department of Philosophy, Xi'an Jiaotong University, Xi'an, China

C. Ding, R. Fan (eds.), *Chinese Ethical and Legal Perspectives on Automated Vehicles*,
https://doi.org/10.1007/978-981-96-1277-2_3

the fate of the people in order to bring self-driving cars to the masses" (JafariNaimi, 2018, p. 303). What is at stake in formulating a moral algorithm is deciding which moral principles or values should be endorsed and programmed in autonomous systems for AVs. However, the debate around this question is easily bogged down in conflicting opinions: Should we buy a car programmed to sacrifice its passengers in some circumstances? Or should we prefer our cars programmed to protect their passengers in all circumstances?

This chapter will address these tricky questions by appealing to some Confucian moral values and related moral principles. It is crucial to incorporate a Confucian perspective into this discourse, not only due to the profound social and cultural integration of technology, but also because it could help shed light on the interaction and relationship between machines and humans.

There has been a great deal of discussion of moral algorithms, but missing from these discussions is a discussion of a deeper, antecedent question: what is the nature and application scope of moral algorithms? The issue of moral decision-making provides an excellent context for us to see the different ways in which machines and humans "do" ethics. And making this distinction clear is essential to building a trustworthy moral algorithm. A Confucian ethical perspective can help us recognize this distinction.

In the following, we initially differentiate crash scenarios involving AVs from the so-called trolley cases to underscore the key considerations in developing an ethical framework for automated Vehicle crashes. It is argued that trolley cases are fundamentally distinct from real-world collisions, which algorithmic morality for AVs is supposed to deal with. Following this line of thought, what is at stake for the moral algorithm of AVs is not a decision procedure for AVs, as an application of the moral rules informed by the trolley cases, but a complete moral account with an emphasis on human-machine interactions. In this chapter, I will elaborate on a Confucian relational view of the ethics of technology, trying to provide a comprehensive moral account of moral decision-making for AVs. After that, based on this Confucian ethical framework as well as these underpinning Confucian moral values, such as harmony and virtue, we will examine the different ways automated vehicles are programmed to behave in the event of an accident and then give an ethical evaluation according to different levels of autonomy for self-driving systems.

Ethical Decision-Making for AVs

Prior to delving into algorithmic morality for automated vehicles, it is essential to examine the distinct manner in which machines navigate ethical decisions. This involves differentiating between the potential crash scenarios that AVs may face in reality and the classic examples often associated with the trolley problem. Although the so-called "trolley problem" has dominated the moral landscape of accident-algorithms, more and more researchers are cautious about the analogy by arguing that the trolley problem in the real world is categorically different from the problem of moral decision-making for AVs.

According to Thompson (1985, p. 544), the trolley problem consists of several similar cases designed to elicit conflicting moral intuitions, while a solution to the trolley problem can explain the moral differences between different cases and inform the underlying moral principle. However, it is hard to uncover the moral principle in a similar way in the context of accidents for self-driving cars, just as automated systems are not merely replicas of human drivers.

Specifically, there are three compelling reasons why it is morally imperative to differentiate between the trolley problem and the crash scenarios that AVs are designed to address.

Firstly, the conventional version of the trolley problem is intentionally crafted to evoke moral intuitions regarding isolated scenarios, prompting individuals to consider their course of action in a given dilemma. In contrast, algorithmic moral decision-making can only be made collectively in abstract terms (Schäffner, 2021). This implies that the moral principles elucidated in the trolley problem, whether viewed through a deontological or utilitarian lens, cannot be directly transposed to the realm of autonomous driving. Humans rely on intuitive decision-making, a capability that machines lack. Some moral reasons humans can recognize intuitively are complex to encode in moral algorithms. JafariNaimi (2018, p. 311) observes that "the algorithm relies on categories such as human/nonhuman, young/old," but "such categorization poses an ethical and political challenge" because "what is included as a category and how it is included can be a source of deeply problematic biases."

Second, the trolley problems, designed as thought experiments, are abstract from the actual ethical situation for AVs in the real world. Two examples: (1) trolley problems leave no room for special obligations. Trolley problems prompt individuals to make moral decisions in hypothetical scenarios, devoid of any specific contextual information. However, in real-life situations, the occupants of a vehicle typically have personal relationships with the passengers, who are likely to be friends or family of the driver, as well as with the pedestrians on the road. It is unrealistic to expect individuals to disregard these personal connections and the unique obligations that come with them when making a moral decision that could have significant consequences for all parties involved. Instead, a rational approach to moral decision-making for automated vehicles should involve a comprehensive understanding of the specific circumstances at hand, a perspective that is inherently obscured by the narrow confines of traditional trolley problems. (2) The trolley problems leave out the considerations about moral and legal responsibilities, which are crucial for formulating ethics for AVs. As Keeling (2020, p. 298) points out: "Trolley cases exclude information about which of the affected parties is responsible for the collision. In real-world collisions, facts about who is responsible in part explain the moral permissibility of the AV's acts."

Lastly, the trolley problem operates under the assumption that there exists a definitive solution capable of reconciling the apparent conflicts between scenarios and thereby elucidating the relevant normative moral principle. For example, Thompson uses it to make the renowned moral difference between the "positive" and "negative" duties.

In contrast, the trolley scenarios specific to the realm of self-driving vehicles inherently contain irresolvable conflicts. Consider, for instance, a scenario where your autonomous car is transporting you and your family towards a tunnel when a child unexpectedly emerges in your path. With a collision inevitable, the situation presents only two options: either the car strikes the pedestrian, resulting in her death, or the car veers into the tunnel wall, leading to the passengers' demise. What should the car choose? Should it choose to protect the passengers at all costs, or should it choose to protect the pedestrian to minimize social harm?

There is no easy way out of the dilemma, in the sense that such a dilemma could be termed the tragically unresolved moral dilemma proposed by Rosalind Hursthouse (1999). The most important reason why we cannot solve this moral dilemma under debate by merely appealing to normative principles is that this dilemma is highly contextualized: the assessment of the action varies with the perspective it is taken from. In other words, the right action depends not only on the concrete situation but also on the different positions different actors might take. On the contrary, the trolley problem is often analyzed in a detached and decontextualized manner, typically from the viewpoint of a disinterested observer. It is important to recognize that this detached perspective is seldom applicable in the context of autonomous driving accidents, as all individuals involved in the collision are stakeholders profoundly impacted by the vehicle's decision.

Similarly, such cash dilemmas for AVs give rise to a social dilemma: "Even though people seem to agree that everyone would be better off if AVs were utilitarian (in the sense of minimizing the number of casualties on the road), they all have a personal incentive to ride in AVs that will protect them at all costs" (Bonnefon et al., 2016, p. 1575). Such a social dilemma vividly reveals that what is at stake in the self-driving dilemmas is not so much conflicting moral principles as conflicting interests and positions of all involved parties. In other words, what seems to be a moral question at first glance is ultimately a political problem. As Himmelreich persuasively argues (2018), the frame of trolley problems locates the problem on the wrong level by providing a moral answer to what ultimately is a political issue. To put it in another way, what seems needed is a social choice about how to overcome disagreements over values and interests, rather than individual moral intuition.

Drawing upon these substantial moral distinctions, it is contended that the limited scope of trolley problems is insufficient for addressing the moral decision-making challenges faced by autonomous vehicles (AVs). Its highly abstract nature is distant from actual collisions in the real world. Furthermore, it masks the notable moral disparity between human and machine ethical decision-making processes. Whereas humans typically make ethical decisions intuitively in the moment, the decisions relevant to vehicles are prospective or contingency plans pre-programmed in a collective manner. As Schäffner (2021, pp. 26–257) correctly states, "A variety of stakeholders such as, e.g., lawyers, philosophers, politicians, and manufacturers are involved in the process of elaborating behavioral responses to dilemma situations in autonomous driving contexts", and the crash algorithms "are expected to take into account the perspectives of all affected parties such as future passengers, other groups of roads users as well as uninvolved persons." These observations suggest that we must take seriously the characteristics of ethical machine decision-making, as well as its impacts on all those involved parts, in order to build ethical autonomous machines.

In sum, consonant with other authors are against putting the problem of ethical decision-making for AVs in terms of the trolley dilemmas

(Nyholm & Smids, 2016; Himmelreich, 2018; Morning & Meschtscherjakov, 2019; Schäffner, 2021), this chapter argues that we need to go beyond the narrow frame of trolley problems and look for a complete moral picture, which could take into account different interests, different positions of stakeholders. In this respect, I believe that Confucian ethical resources can provide a highly illuminating perspective by conducing to maintain a dynamic and balanced moral relationship between humans and machines.

What We Could Learn from Confucian Ethics

In proposing an alternative Confucian ethical framework, we need to know more about how Confucianism deals with moral dilemmas and what important ethical insights it will bring to our resolution of moral dilemmas in the context of autonomous driving. Although we cannot find cases of life-and-death decisions like the trolley dilemmas in Confucian classics, there is no lack of discussion about moral dilemmas. A typical case is whether the children should denounce their father if he has stolen a sheep. In terms of law and order, the children should denounce the father. However, from the point of view of father-son bonding, it seems that the children should not denounce their father. The Duke of Sheh informed Confucius of this case and said, "Among us here, there are those who may be styled upright in their conduct. If their father has stolen a sheep, they will bear witness to the fact." But Confucius proposed a different approach to this ethical dilemma. In his reply to the duck, he said, "Among us, in our part of the country, those who are upright are different from this. The father conceals the misconduct of the son, and the son conceals the misconduct of the father." (Analects, 13) As Zhao Tingyang (2015) astutely points out, here Confucius is not arguing in favor of egoism, as if to give precedence to "selfishness," but rather to go beyond conflicting ethical norms to reveal the deeper moral grammar behind the norms. A deeper moral grammar is one in which the "normative" rules of ethics are underlain by "grammatical rules" as a deeper structure, embodying a robust understanding of how to live a life. Such a deep moral grammar cannot be reduced to some normative rules; instead, it determines at a deeper level the specific scope of application of the moral norms. In the case under discussion, the general and genuine ways humans are related to each other is the profound moral grammar to which Confucius appeals. "The love between father and son is the universal human condition, and if fathers and sons betray each other, it is detrimental to familial relations and deteriorates the universal human condition. People usually believe that the relationship between father and son is the most intimate and trusted. Without loyalty between father and son, people will quickly lose trust in each other, and society will lose the basis for its harmonious functioning" (Zhao, 2015, p. 102).

This Confucian approach to moral dilemmas echoes similar insights from virtue ethics. Hursthouse (1999), from a perspective of virtue ethics, notes that there are "intractable moral dilemmas," or more precisely, moral dilemmas that cannot be resolved consistently based on readily available moral norms. Various moral agents may arrive at different moral conclusions when faced with the same ethical dilemma, and such discrepancies are considered morally acceptable. Confucianism aligns with the fundamental notion that moral standards cannot be rigidly imposed in abstract terms, detached from real-world contexts and situations.

Rather than viewing moral dilemmas as problems to be resolved, Confucianism tends to view moral dilemmas as the starting point for moral reasoning. What is at stake is not so much how to resolve moral dilemmas in a standardized way but how to balance disturbed ethical relationships, which is very much dependent on the decision-maker's situated knowledge of the decision-making context. In this regard, representing moral dilemmas as conflicting ethical norms somewhat obscures the real focus of the matter.

The above Confucian insight may help us to explain why the rational approach to algorithmic morality is unsatisfactory. In the following section, we will try to explain the moral defects of the main algorithmic ethical strategies from a Confucian perspective. We will argue that the common flaw of both utilitarian and deontological strategies is the attempt to reduce the problem of ethical decision-making in autonomous driving to a problem of normative choice: i.e., what kind of moral norms should we base our programming of the machine on, or what kind of socially acceptable ethical norms should the machine act on. Meanwhile, the more profound question is obscured or marginalized: Should we let machines make decisions instead of people? What are the key differences between machine decision-making and human decision-making? What are the ethical implications of these critical differences, not only in terms of algorithmic ethics but also in terms of human flourishing?

The primary ethical concern often associated with utilitarian algorithmic strategies is that attempting to systematize the ethical foundations of human moral decision-making can lead to ethical problems, such as discrimination. Imagine a self-driving car faced with the difficult choice of turning right and running down an eight-year-old girl or turning left but tragically taking the life of an eighty-year-old man (Lin, 2016). A utilitarian ethical perspective would favor saving the little girl, though at the price of the old lady's life. Moreover, the utilitarian justification for this choice could be expressed touchingly from a human point of view: "The girl still has her entire life in front of her-a first love, a family of her own, a career, and other adventures and happiness-while the grandmother has already had a full life and her fair share of experiences" (Lin, 2016). However, the algorithms tend to see the crashing situations from a machine's point of view, which can be very different from a human view. As mentioned earlier, machine decisions are prospectively made in a collective way, and thus, the algorithms' generalization of the utilitarian rationale mentioned above can have unintended adverse effects, such as discrimination against the elderly. Justifying such discrimination ethically poses a significant challenge, underscoring the ethical distinctions between human and machine decision-making processes.

Another ethical flaw of utilitarian strategies is also related to the different ethical roles of humans and machines. Imagine a scenario where a self-driving car that is driving you on a narrow road is about to collide with a school bus full of children. Moreover, the car calculates that no possible action can avoid harming you. What should your car do?

Given that there is, your car seems to have and only have two choices: (1) to crash into the bus, risking everyone's lives, or (2) to drive off the cliff, sparing the lives of everyone on the bus. An autonomous car programmed with a consequentialist framework would seem to choose to sacrifice the passenger. However, as Patrick Lin points out, a consequentialist framework fails to recognize a crucial ethical distinction between humans and machines. "It is one thing for you to willingly make the decision of sacrifice yourself, and quite another matter for a machine to make that decision without your consent or foreknowledge that self-sacrifice was even a possibility" since such a decision is "one of the most important that can be made about one's life-perhaps noble if voluntary, but criminal if not." This contrast vividly highlights the fundamental disparity between the functions fulfilled by humans and machines in ethical decision-making. It is crucial to acknowledge and differentiate the distinct moral responsibilities of humans and machines, as failing to do so may give rise to significant ethical concerns and dilute the genuine essence of the ethical choices at hand. Only through complete autonomy can the ethical value of self-sacrificial decisions be ensured; this necessitates that the ultimate decision-making authority always rests with the human driver, rather than the machine.

The greatest challenge in constructing a deontological ethical framework for Autonomous Vehicles (AVs) lies in resolving potential conflicts between ethical norms. Determining the hierarchy of rules in a given situation demands human-like common sense and contextual understanding, qualities that machines currently lack. Furthermore, the impartial nature inherent in deontology overlooks the unique moral obligations associated with specific relationships. For instance, in the realm of autonomous driving, the driver or owner of a vehicle typically holds a greater duty of care towards the passengers than towards pedestrians or other individuals in the vehicle. Could we translate this partiality to algorithmic morality in a universal way?

These difficulties bring us back to the original point of the discussion, revealing a more profound but less discussed question: Should we let machines make moral decisions? Should we allow machines to make life-and-death choices instead of people? Do we fully understand the consequences of letting machines make moral decisions? Neither the utilitarian nor the deontological path can take seriously the above questions. Instead, within the ethical frame of trolley problems, both treat machines as human-like moral subjects, as if we could directly appeal to moral rules to create moral machines. However, morality is not simply the observance of rules; the whole meaning of morality has to be rooted in a life that cannot be fully normalized. The following might be the most critical lesson from Confucian ethics: life reveals ethics, and ethics is embedded in life.

As for the ethical context of self-driving collisions, from a Confucian view, the critical issue is to reevaluate the ethical relationships between stakeholders in the decision-making process, which the introduction of autonomous machines has significantly changed. A more reasonable goal for formulating an ethical framework for autonomous driving is not to build human-like moral machines but to set expectations about what we could expect from an autonomous machine. To do so, we need to develop an ethical framework that can both accommodate the potentially divergent positions of different participants and the shifting decision-making scenarios and, at the same time, be complete and resilient to grasp the complex and delicate ethical situation in the context of self-driving accidents. In this paper, we will draw on Confucian ethical resources to propose a relational technological ethical framework to advance the relevant explorations. Specifically, this framework is role-based, emphasizes harmony norms, and centers on virtues in the realm of technology ethics.

A Confucian Relational View of Machine Ethics

The Confucian role ethics proposed by Roger T. Ames and the Confucian civility of Jeol J Kupperman (2010) have both influenced my account here. Pak-Hang Wong's (2010) paper 'Dao, Harmony and Personality: Constructing a Confucian Essay on Technology' is one of the first studies to attempt to advance in this area. Like Kupperman, Wong, and other authors, the author argues that the relational perspective of Confucian ethics has ethical concerns and a moral focus that is quite different from the mainstream Western ethical framework. However, this different perspective can stimulate intellectual reflection and rehabilitate itself to help us construct a more comprehensive and balanced ethical framework to address the complex ethical challenges of the real world.

Although the focus of different interpreters may vary, a relational Confucian ethical perspective will always resort to three fundamental concepts or values: role, harmony, and virtue. In the following section, we will briefly elaborate on the fundamental ethical connotations of these three concepts and then examine their moral implications for automated vehicle ethics from a relational Confucian ethical perspective.

The role itself is a relational concept, and it is challenging to discuss its connotations outside of the context of a particular relationship. Although we may not find a strict equivalent to the concept of role in the Confucian canon, there is a clear connotation of role in the 'proper name' (正名) in the Analects. In Confucius, what is associated with the names 'father' and 'son' are no longer isolated people and events but the intricate network of social, cultural, and life relations. People and events are constructed and sustained through the network lying behind these names.

These Cultural and life-related networks, contexts, and forms make a social role simultaneously nominal and possess the power to claim moral obligation (Wang Qingjie, 2016, p. 168) Ames's proposal of Confucian role ethics is based on the same insight, namely that the Confucian conception of ethical life can ultimately be traced back to the recognition of the fact of human relational existence: 'We are not alive as a mere physical being; we do Everything we do, physical, psychological and social, is unquestionably relational and collaborative" (Ames, 2017, p. 1). Roles as the starting point of Confucian ethics are, on the one hand, representations of various relational patterns, presented as various social roles that support daily life, such as parents, children, grandparents, friends, and neighbors; on the other hand, these roles connote ethical normativity (normative), indicating to us the way to behave appropriately. "Once relationships exist, the prosperity of the family and the state is the best we can achieve according to the conditions of these relationships" (Ames, 2017, p. 1). In this regard, Confucian roles are also known as ethics of flourishing, which is very close to the ethical concerns of Western virtue ethics. However, there are some subtle differences between Confucian role ethics and virtue ethics. We will return to this issue in our explanation of the value of Confucian harmony.

It is essential to emphasize that technology should be viewed through the lens of role ethics, recognizing that its role goes beyond mere instrumentality. Nevertheless, the 'role of technology' remains an important and open question without any definitive and dominant answer. Alternatively, when technology is seen as a mere tool, the most profound ways it affects human life are obscured. We can consider the role of technology in at least two ways: first, by looking at the relationship between humans and the world and examining how intelligent machines have changed the way humans practice, whether they empower humans, improve the human condition, or whether they lead to a deeper disempowerment of human beings. Second, in terms of human relationships, role ethics investigates how intelligent machines have influenced and redefined ethical connections among humans. It also considers whether these changes are consistent with a human vision of the good life.

From the perspective of role ethics, the central question for the ethics of autonomous driving is: What roles should be assigned to intelligent machines?

As mentioned above, all questions concerning roles are posed in a relational context, and thus, what echoes here is also the most fundamental question in the age of intelligent machines: What is the human use of human beings? Yuk Hui (2023) formulates this question as a crisis of human identity of our time, "In the age of the information machine or the cybernetic machine, the machine itself becomes the organizer of information, and human beings are no longer at the center of it, even if they still think of themselves as the commanders of the machine and the organizers of information. This is the moment when humans suffer from their stereotypical beliefs about machines: they mistakenly designate themselves as the center, and in doing so face a constant frustration and a panicked search for identity." The problematic realization of this crisis is even more salient in the context of autonomous driving: Should we allow machines to decide who lives and who dies?

Confucian role ethics offers valuable insights into this complex issue. First, Confucianism would give an apparent negative answer to the above question and emphasize that humans and machines play different roles depending on their natures. The fact that ethical rules (utilitarian or deontological) cannot be fully translated into algorithmic suggests that humans and machines can never act in similar ways. Creating moral machines that act precisely in the same way as humans is indulging in fantasy. For Confucians, proper moral reasoning must begin with an awareness of the range of different roles of man and machine and the specific ways of their interactions.

Secondly, from the perspective of responsibility, AI cannot be a responsible moral agent in the complete sense of the word. Thus, we should attribute the ultimate decision-making power to humans. There are three ways to imagine the moral agent role that AVs could play, i.e., the ethical-impact agent, implicit ethical agent, and explicit ethical agent (Moor, 2006). When some technology can be valued in ethical terms and impact ethical issues such as privacy and property, we can call it an ethical impact agent. However, such an ethical impact agent is not yet an ethical actor in the true sense of the word insofar as it does not behave internally ethically. The latter two align with what it means to be an ethical actor. According to Moor, a machine as an implicit ethical agent would act ethically because its internal functions implicitly promote ethical behavior. An airplane's automatic pilot is a good example of implicit ethical agents, which vividly shows that implicit ethical agents should always be under human agents' control and serve as instruments for fulfilling human purposes.

In contrast, explicit ethical agents are capable of independently making moral decisions, or as Moor (2006, p. 16) put it, "can 'do' ethics as the computer can play chess." Confucianism will likely view AVs as implicit ethical agents because they fall short of autonomous moral responsibilities. For Confucians, the existential difference in moral responsibility between humans and machines is a red line that cannot be crossed. When the boundary is blurred, people will have to face some morally troubling scenarios: for example, the self-driving car you are traveling in decides to sacrifice you, whether you like it or not.

Following this line of thought, Confucians would argue that the primary virtue of a machine is "loyalty." As Liu (2018) points out, loyalty to one's task is crucial for formulating Confucian machine ethics because AI should particularly be designed to be role-specific. Autonomous vehicles are permanently prohibited from making decisions that sacrifice the vehicle's passengers, even if, in some cases, such sacrifices could lead to greater public good. Only human drivers in autonomous vehicles can make self-sacrificing decisions because only human agents can make autonomous moral decisions.

The second key conception of Confucian machine ethics is harmony. At the heart of the Confucian way of prosperity is harmony, as the first chapter of *The Doctrine of the Mean* states: "While there are no stirrings of pleasure, anger, sorrow, or joy, the mind may be said to be in the state of equilibrium. When those feelings have been stirred, and they act in their due degree, there ensues what may be called the state of harmony. This equilibrium is the great root from which all the human acts in the world, and this harmony is the universal path that they all should pursue. Let the states of equilibrium and harmony exist in perfection, and a happy order will prevail throughout heaven and earth, and all things will be nourished and flourish." Here, harmony (和) identifies the Confucian understanding of the ultimate order of the world. This order, which cannot be reduced to the supremacy of some entity beyond the world, opens up a realm of life within the world, in which all things move and form their reason, yet in parallel and without contradiction, merging into a whole. Qian Mu has a vivid metaphor for this: For example, on a big road, there are car lanes, tram lanes, and pedestrian lanes, each following its path, not hindering each other, but still united as one "Dao". … The birds fly, and the fishes leap, that is, the whole freedom of the lively world (Qian Mu, 2012, pp. 41–42). In the words of the roles above, all things, such as cars and trams, have their way, and the combined way is not outside all things but exists among them and is manifested as a harmonious relationship between the roles. In this sense, the way is harmony; that is to say, as *The Doctrine of the Mean* put it, "All things are nourished together without their injuring one another."

Harmony, as a cardinal ethical value, has three essential features that deserve attention. First, as Li Chenyang has already pointed out, harmony is the normative standard of Confucian ethics. Confucian ethical principles cannot usually be expressed as a decontextualized universal rule, such as the Kantian universal obligation not to lie. However, this does not mean that Confucian ethics does not provide rules for the right actions. For Confucians, harmony is the most crucial criterion.

The second characteristic of harmony is that it is often a search for a balance between the opposites, i.e., the pursuit of 'harmony but difference.' Moreover, this 'sameness in difference' is not created by some unprincipled compromise but rather by the 'opposite.' Opposites can mutually enhance each other in a harmonious relationship. Thirdly, regarding maintaining harmony, Confucianism stresses that harmony can only be a dynamic balance and that there is no predetermined order. Furthermore, humans are given a prominent role in maintaining harmonious relationships. As Mencius said, "All things are already complete in us. There is no greater delight than to be conscious of sincerity on self-examination." (*Mencius*) Humans have the primary moral responsibility for maintaining harmony in themselves and everything. The Confucian concept of harmony has its origins in a particular insight into the complexities of real life; it is perhaps out of this profound awareness of the complexities and subtleties of practical ethical life that the Confucian tradition does not seek an absolute, decontextualized norms of 'goodness' or 'rightness.' This insight also makes the Confucian ethical framework more inclusive and accommodative of new technological developments than competing ethical approaches. For example, Confucian ethics and Aristotelian virtue ethics have overlapping ethical horizons, since they share the same commitment to interpreting how humans flourish. One key difference, however, is that the ultimate good sought by Confucianism (e.g., harmony) is more formalized and tends to be expressed more as a regulative ideal of dynamic processes. In this respect, the Confucian ethical framework is more flexible than virtue ethics in the general sense. The moral foundation of virtue ethics typically starts with a teleological discussion on human flourishing. However, when this teleological perspective is detached from a shared context of common understanding, reaching a consensus becomes challenging and may lead to relativism. In contrast, the Confucian principle of harmony can be comprehended and upheld without relying on a detailed teleological framework of human flourishing.

The Confucian concept of harmony focuses on the dynamic and balanced interactions among various individuals, rather than on universal moral principles. Harmony is the natural result of interconnected actors playing their own roles. Moral failure or moral chaos often results from actors acting in ways contrary to the requirements of their roles. Many of the dilemmas in the current ethical discussion of autonomous driving arise from demanding too much of the role of the machine beyond what it can and should assume or arise from misaligned role relationships. For example, Confucians will see it as a typically misplaced role to let the machine decide who it would kill in case of an ethical dilemma.

Moreover, as previously discussed, the harmonious ethical framework places humans at the center, emphasizing their unique and pivotal role distinct from all other beings. In the context of moral decision-making for AVs, a Confucian ethical and moral picture would, at most, accept the machine as an implicit ethical agent, supporting rather than replacing human moral agents. Confucian notions of harmony would require the development of a model of human-machine co-decision-making, putting machine behaviors under meaningful human control. This model is inherently pluralistic. We can envisage a three-phase decision-making model.[1] In the first phase, some universal moral rules are built into the algorithms, serving as the behaviors of the machines. "Rules should consist of widely agreed upon concepts, e.g., injuries are preferable to death, property damage is preferable to injury, and vulnerable users should be protected foremost" (Goodall, 2014). In the second phase, machines can use machine learning techniques to understand the correct ethical decisions and manage risks. The third phase consists mainly of human-machine communication and interaction, preferably using natural language, aiming to hold the machine under meaningful human control.

To conclude, through the lens of harmony, Confucian machine ethics pay more attention to the relation and interactions between humans and machines, concerned with the technology for the nontechnological parts of the world, rather than merely building an explicit and full-fledged artificial ethical agent in its own right. This relational and harmonious approach expands the scope of research into moral decision-making for AVs.

The quest for harmony always comes down to a concern for virtue. On the one hand, harmony is always founded upon the corresponding virtues exhibited by the actors in the interpersonal relationship. A considerable body of research has pointed out that Confucian ethics shares similar

[1] Goodall's proposal influences the model discussed here. However, a key difference exists between the Confucian version and Goodall's model. Confucianism argues that the machine can only play a supporting role in the decision-making process. However, Goodall argues that the latter can be analogized to a child, hoping that one day he could reach a higher state, i.e., to be a full-fledged ethical agent (Goodall, 2014).

concepts and logic with virtue ethics, which originated in ancient Greece, i.e., a virtue-based framework of moral reasoning. Just as a virtue in Aristotle is 'the state that makes a person good and enables him to fulfill his function,' so too does Confucian 'virtue' have the connotation of humane fulfillment.

On the other hand, Conversely, when harmony is employed as a normative criterion for evaluating relationships, virtue also holds significant importance. If introducing technology into a human practice and the consequent reconfiguration of the roles involved results in the erosion of the moral actors' virtue, then we need to think carefully about the ethical applicability of this technology. Furthermore, the unique sensitivity of the Confucian virtue ethics perspective to interpersonal relationships may also shed light on specific values that are obscured in other ethical frameworks. This unique moral perspective explains why it is essential to resort to Confucian thought in the ethics of technology, not merely because of our cultural preference but also because of a cross-cultural and universal ethical imperative.

From such a virtue perspective, Confucian ethics does not believe that the human desire for comfort and efficiency provides sufficient moral justification for developing driverless cars. Whether and what kind of autonomous driving technology should be developed needs to be valued from an agent-centered perspective of virtue ethics.

From the perspective of virtue ethics, Confucianism would argue that "the ethical knob" (Contissa et al., 2017) is an ethically problematic device. Although the "ethical knob" can help the public accept self-driving cars by ensuring autonomous choice, the users very easily fall into the prisoner's dilemma, which would hinder the cultivation of virtue that Confucianism values. Therefore, in the Confucian view, the implementation of "the ethical knob" is morally problematic because it essentially encourages people to become more egoistic rather than altruistic. Neither a higher insurance premium nor a more limited coverage can compensate for this moral loss.

How to Design Algorithmic Morality for AVS: A Confucian Account

From the perspective of Confucian role ethics, it is important to distinguish between the role of the machine and the human being. Firstly, machines cannot have the full-fledged moral agency that humans have. Secondly, machine decision-making is insufficient to deal with the complexity of self-driving decision-making situations. Finally, it is a degradation of human dignity to put machines in the position of making decisions that could have life-and-death implications.

Furthermore, Confucian ethics argue that confusing the role of machines and the role of humans in the process of moral decision-making is a significant moral failure. We can use an example to illustrate the significant moral difference between the role of machines and humans. Suppose I am driving a self-driving car on a mountain road, and suddenly, a school bus full of children comes up to me and says that a severe accident is about to happen. If, at this point, I autonomously decide to drive towards the cliff and avoid harming the children on the school bus, then my actions would undoubtedly be a heroic moral feat. Nevertheless, if the machine decides to sacrifice me without consulting me, then the whole nature of the act changes, even if the outcome is the same. Even if I also have the will to self-sacrifice, this decision must be made by a human being to ensure the full moral content. How decisions are made is of great importance here. For these reasons, according to Confucianism, machines should only play a supporting role in the process of ethical decision-making for AVs, and humans should always retain the final decision. In other words, the function of intelligent machines is to help humans make better decisions, not to replace them. This means that we must keep the steering wheel for self-driving cars, i.e., always keep the final decision-making in the hands of humans. It also means that we must think carefully about the scope of use of autonomous vehicles.

As we discussed above, Confucianism views harmony as the criterion of rightness. Let us return to the questions raised at the beginning of this chapter: Should we buy a car programmed to sacrifice its passengers in some circumstances? Or should we prefer our cars programmed to protect their passengers in all circumstances? The Confucian answer to both questions is no because neither the "harm minimization" policy nor the "passenger first" policy meets the criteria for harmony. The "harm minimization" policy is insufficient to address the crash situation adequately, as minimizing harm is not the same in all circumstances. Simple applications of the "harm minimization" policy sometimes yield unfair outcomes. For example, the "harm minimization" policy may require AVs to choose to hit a person wearing a helmet rather than a person not wearing one, even though the latter violates traffic rules. The "passenger first" policy is also ethically problematic. Following the rule of protecting the passenger at all costs, the AVs will cause disproportionate harm to other road users. Just as harmony is not some pre-existing order, neither can an autonomous system always produce harmonious outcomes by mindlessly following pre-implemented specifications. To sustain a harmonious relationship among all involved parties, we should ensure automated vehicles are ultimately under the control of some responsible human agent.

Depending on the different autonomous levels of AVs, harmony can also be expressed in different ways. For levels 1 to 3 autonomous driving, the role of the autonomous driving system should be to facilitate and improve the ability of humans to make ethical decisions rather than replace them.

Furthermore, envisioning another driverless harmony for fully automated vehicles is also feasible. However, the prerequisite for this driverless harmony is that the data is shared sufficiently so that the system can coordinate the reactions of the autonomous drivers.

Virtue is an essential lens through which we understand and evaluate technology. Regarding virtue, autonomous driving technology's greatest challenge is the undesirable "responsibility gap" or "accountability vacuum." On the one hand, delegation of decision-making to an autonomous system diminishes human moral responsibility over decisions and actions. In contrast, a sense of inherent responsibility is integral to human moral agency. On the other hand, the introduction of machine decision-making. Although there is no easy way to address these challenges, a virtue perspective can broaden our moral horizons. One of the crucial points is that we must consider both tangible and intangible effects, such as how this autonomous system might change the internal character of human moral agents.

The Confucian frame discussed above contributes to AVs' moral design and regulation in the following ways. By distinguishing between the different roles of humans and machines, this Confucian frame advocates that responsible human agents must control Level 1–3 automated vehicles. Fully automated vehicles can only be used on exclusive, closed roads specifically built for driverless cars. Moreover, the new technology should be perceived and evaluated in terms of virtues, which are inherently difficult to standardize and translate into algorithms.

Concluding Remarks

From the perspective of Confucian machine ethics, this chapter introduces a framework centered on meaningful human control to shape the ethics of AVs. This framework shifts the ethical focus from creating a distinct artificial moral agent to fostering dynamic and balanced relationships between humans and machines, with the goal of enhancing the moral decision-making of human operators. In the Confucian view, algorithmic morality should be based on human-machine harmonious interaction and relationship rather than unthinkingly following pre-implanted, ready-made rules. In other words, the problem of ethical decision-making in the context of autonomous driving cannot be reduced to ethical algorithms or algorithmic ethics. Instead, the rationality of algorithmic ethics is premised on the harmonious interaction between humans and machines. Many ethical resources in Confucianism deserve further exploration in this regard. As an extensive discussion of this question is beyond the scope of this chapter, just some tentative thoughts, such as role, harmony, and virtue, are sketched at this point.

Acknowledgements This research is supported by the National Social Science Fund of China (grant no. 20BZX127).

References

Ames, R. T. (2017). *Confucian role ethics: Vocabulary*. The Chinese University of Hong Kong Press.

Bonnefon, J., Shariff, A., & Rahwan, I. (2016). The social dilemma of autonomous vehicles. *Science, 365*, 1573–1576.

Contissa, G., Lagioia, F., & Sartor, G. (2017). The ethical knob: Ethically-customisable automated vehicles and the law. *Artificial Intelligence and Law, 25*, 365–378.

Goodall, N. (2014). Ethical decision making during automated vehicle crashes. *Transportation Research Record: Journal of the Transportation Research Board, No. 2424*. Transportation Research Board of the National Academies, Washington, DC, 58–65.

Himmelreich, J. (2018). Never mind the trolley: The ethics of autonomous vehicles in mundane situations. *Ethical Theory and Moral Practice, 21*(3), 669–684.
Hui, Y. (2023). ChatGPT, or the eschatology of machines. *E-flux Journal, 137*, 1–8.
Hursthouse, R. (1999). *On virtue ethics*. Oxford University Press.
JafariNaimi, N. (2018). Our bodies in the trolley's path, or why self-driving cars must not be programmed to kill. *Science, Technology & Human Values, 43*(2), 302–323.
Keeling, G. (2020). Why trolley problems matter for the ethics of automated vehicles. *Science and Engineering Ethics, 26*, 293–307.
Kupperman, J. J. (2010). Confucian civility. *Dao, 9*, 11–23.
Lin, P. (2016). Why ethics matters for autonomous cars. In J. M. Maurer et al. (Eds.), *Autonomous driving: Technical, legal and social aspects*. Springer.
Liu, J. 刘纪璐. (2018). Confucian robotic ethics (儒家机器人伦理). *Ideas and Cultures, 1, 18–40.*
Moor, J. (2006). The nature, importance, and difficulty of machine ethics. *IEEE Intelligent Systems, 21*(4), 18–21.
Morning, A., & Meschtscherjakov, A. (2019). Trolled by the trolley problem: On what matters for ethical decision making in automated vehicles, *CHI '19: Proceedings of the 2019 CHI Conference on Human Factors in Computing Systems*, Paper No. 509, 1–10. https://doi.org/10.1145/3290605.3300739
Nyholm, S., & Smids, J. (2016). The ethics of accident-algorithms for self-driving cars: An applied trolley problem? *Ethical Theory and Moral Practice, 19*(5), 1275–1289.
Qian, M. 钱穆. (2012). *The record of idle thoughts on the lake* (湖上闲思录). Jiuzhou Publishing House.
Schäffner, V. (2021). Between real world and thought experiment: Framing moral decision-making in self-driving car dilemmas. *Humanistic Management Journal, 6*, 249–272.
Thompson, J. (1985). The trolley problem. *The Yale Law Journal, 94*, 1395–1415.
Wang, Q. 王庆节. (2016). *Moral moving and Confucian demonstration ethics* (道德感动与儒家示范伦理学). Peking University Press.
Wong, P.-H. (2010). Dao, harmony, and personality: Towards a Confucian ethics of technology. *Philosophy & Technology, 25*(1), 67–86.
Zhao, T. 赵汀阳. (2015). The forking paths for the trolley problem (有轨电车的道德分叉). *Philosophy Research, 5*, 96–102.

CHAPTER 4

"I Must Have a Steering Wheel in My Automated Car to Protect My Family": A Confucian Ethical Reflection on Programming Automated Vehicles

Ruiping Fan

Introduction

The emergence of automated vehicles (AVs) has raised significant ethical, legal, and social issues. This chapter aims to explore the ethical considerations surrounding the design of AVs in response to inevitable accident scenarios from a Confucian virtue ethical perspective. The focus is on suggesting appropriate software programming for AVs to be used in contemporary Confucian-influenced societies like Hong Kong and mainland China.

This chapter does not aim to provide a comprehensive Confucian ethical code for programming and managing AVs. Instead, it concentrates

R. Fan (✉)
Department of Public and International Affairs, City University of Hong Kong, Kowloon, Hong Kong
e-mail: safan@cityu.edu.hk

C. Ding, R. Fan (eds.), *Chinese Ethical and Legal Perspectives on Automated Vehicles*,
https://doi.org/10.1007/978-981-96-1277-2_4

solely on proposing specific ethical guidelines for deciding the digital algorithms of AVs and determining their course of action when faced with "brake failure scenarios"—situations where stopping before hitting human pedestrians or bystanders on the road is deemed impossible.[1] The central question is whether an AV should be programmed to prioritize the safety of its occupants by staying on course and hitting bystanders or to swerve and risk harm to its occupants in order to protect the lives of external individuals. While such scenarios may only arise infrequently once AV technology is sufficiently advanced, they hold immense importance and are of great concern to potential AV users. Hence, the programming choices made for handling such circumstances will significantly influence the willingness of East Asian individuals to own or utilize AVs.

In the following, section "Culture-Irrelevant Approaches" presents an overview of the prominent culture-irrelevant approaches to AV ethics found in the existing literature. These include consequentialist and self-protective approaches, a deontological perspective, an Aristotelian virtue-ethics perspective, and a hybrid approach. Section "Towards a Culture-Relevant Approach" examines the limitations of each of these comprehensive universalist approaches, highlighting their general shortcomings and incompatibility with the moral culture of Confucian regions specifically. This section also provides a concise summary of key features of Confucian virtue ethics, which serves as the dominant moral framework in these regions, and briefly responds to a few challenges that would be posed on my Confucian culture-relevant approach. In section "Confucian AVs", I propose specific recommendations rooted in Confucian virtue ethical methods and considerations for programming AVs in brake failure scenarios, arguing that different types of AVs for various uses should be programmed in different ways. I offer justifications for these suggestions by drawing upon the ethical resources provided by Confucian moral culture. Finally, section "Concluding Remarks" offers insights on why it is unreasonable to expect all regions with diverse cultures to adopt the exact same AV algorithms to address brake failure situations.

[1] When referring to "brake failure scenarios" in this chapter, I intend to encompass all situations involving unavoidable collisions or crashes, regardless of the brakes functioning properly. It is important to note that this chapter does not specifically distinguish between incidents caused by actual brake malfunctions, where the brakes are non-functional, and scenarios arising from the sudden presence of individuals in front of an AV, resulting in the vehicle's inability to stop due to insufficient distance.

Culture-Irrelevant Approaches

This section presents several notable ethical approaches to AVs that have been discussed in the literature. I refer to them as "culture-irrelevant" approaches because they aim to provide universal solutions to ethical issues related to AVs, without taking into account the cultural differences across various regions. In section "Towards a Culture-Relevant Approach", we will examine why these approaches pose general problems and are particularly unsuitable for adoption in East Asian regions.

Consequentialist and Self-Protective Approaches

Consequentialist ethical theories assess the morality of decisions or actions based on their outcomes. Utilitarianism, as the prominent form of consequentialism, evaluates consequences by considering the balance of pleasures and pains they entail. Utilitarians argue that a decision or action is morally right if it produces at least as much overall net pleasures (pleasures minus pains) in its consequences compared to any other alternative for all affected individuals. Consequently, adhering to this utilitarian framework, AVs should be programmed to address brake failure circumstances in a manner that prioritizes protecting or minimizing harm to human lives. This approach, according to the utilitarian understanding, represents the correct way of evaluating the consequences of AVs' programmed actions. Therefore, I propose the term "utilitarian AVs" (UAVs) to describe the AVs programmed in this manner.

UAVs are designed to make decisions that prioritize saving more lives or causing fewer deaths in brake failure scenarios, regardless of whose lives are at stake. For instance, if a UAV encounters two pedestrians in front of it while carrying only one occupant, it should "choose" to swerve onto a barrier, sacrificing the safety of the occupant to spare the lives of the two pedestrians. This decision should be preprogrammed into the UAVs' software to seek the best overall consequences for all parties involved. Of course, following the utilitarian principle, it is reasonable for UAVs to remain on course and protect the passengers when the number of internal passengers and external pedestrians is the same, as swerving onto a barrier would not yield any further utility gain.

However, despite the potential superiority of UAVs in terms of safety and comfort compared to human-driven vehicles, their widespread adoption in the market faces challenges due to individuals' self-protective interests. Interestingly, research has revealed that many people consider it morally right to develop UAVs, which involves programming AVs to minimize casualties in emergencies. However, when asked if they would personally purchase or rent UAVs, their enthusiasm diminishes. Instead, they tend to prefer a self-protective AV model that prioritizes their own safety, even though such a model contradicts the utilitarian principle of maximizing overall happiness they seem to endorse (Bennefon et al., 2016, p. 1574). For the purpose of clarity, let's refer to the AVs programmed to cater to these self-protective interests as "selfish AVs" (SAVs). We seem to confront a discordant reality where individuals recognize the ethical superiority of UAVs but expect others to adopt them while personally opting for SAVs to safeguard themselves. How should we address this discrepancy between people's moral beliefs aligning with utilitarianism and their moral motivations driven by egoism?

Some may think it justifiable for governments to enforce regulations mandating that all AVs be designed exclusively as UAVs rather than SAVs for individual purchase, ownership, or rental. Such regulations would align with people's moral beliefs and, if effectively implemented, objectively maximize overall societal happiness. However, given the aforementioned divergence between moral beliefs and motivations, successfully implementing this regulation becomes challenging. If only UAVs are available and SAVs are eliminated, most individuals would likely refrain from purchasing AVs altogether, opting to continue using their traditional human-driven vehicles. Consequently, the lives saved by UAVs, used by only a small fraction of the population, would be overshadowed by the deaths caused by the delay in widespread adoption and the persistence of human-driven vehicles (Bennefon et al., 2016, p. 1576; Othman, 2021).

Undoubtedly, the continued use of human-driven vehicles poses significant risks, leading to numerous damages and casualties each year. According to reports from the World Health Organization (WHO), road traffic injuries rank as the primary cause of death for children and young adults aged 5–29, resulting in approximately 1.19 million fatalities annually. Moreover, road traffic accidents cost most countries 3% of their gross domestic product (WHO, 2023). Many of these accidents are attributable to human drivers' shortcomings, such as slow reactions, drunk driving, distracted driving (e.g., texting), fatigue, emotional behavior, speeding, and tailgating. These issues can be significantly mitigated or eliminated by replacing human-driven vehicles with AVs. Therefore, even if we were to simply replace human-driven vehicles with SAVs instead of UAVs, it would still yield superior outcomes, substantially reducing traffic accidents, injuries, and fatalities caused by human drivers, while brake failure incidents in SAVs occur only rarely.

Should AVs be programmed as SAVs when dealing with brake failure situations to ensure market success? If the majority of individuals prefer SAVs over UAVs, allowing SAVs would result in an increase in overall net pleasure, albeit not to the same extent as UAVs. Considering the potential challenges in successfully marketing UAVs, SAVs become the second-best choice for consequentialist calculations. Moreover, as long as the lives of users are not endangered, SAVs can still be programmed to achieve better outcomes for all other individuals involved. For instance, in a brake failure scenario, although an AV should not be programmed to swerve onto a barrier endangering the user's life, it should be programmed to turn away from three pedestrians on one lane and move to two pedestrians on another lane to minimize external harm. Accordingly, for those who prioritize making consequentialist considerations when designing AVs, following customers' preferences and designing SAVs would turn out to be right.

This new consequentialist approach, differing from the utilitarian one, appears to have been embraced by the online experimental "Moral Machine" platform developed by the MIT lab. This platform serves to explore the moral dilemmas faced by AVs by gathering input from a majority of individuals. The platform offers "serious games" designed to collect extensive data on how individuals would want or expect AVs to behave in ethical quandaries, thereby providing insight into the preferences of the general population. Since its inception, the platform has provided online dilemmatic games in ten different languages and has amassed 40 million decisions from respondents in 233 countries as of 2018 (MIT, 2022). In summary, the platform has yielded three key findings: people tend to prioritize the preservation of human lives over animals, prioritize saving a greater number of lives over a smaller number, and prioritize saving the lives of young individuals over older ones (Awad et al., 2018).[2]

[2] It is important to highlight a notable cultural distinction observed by the authors of the aforementioned paper. They report that the inclination to prioritize the preservation of younger characters over older ones is considerably less pronounced in countries belonging to the Eastern cluster, which encompasses Far Eastern countries aligned with the Confucian cultural groups and Islamic countries. Conversely, this preference is significantly higher in countries within the Southern cluster, comprising Latin American countries (Awad et al., 2018, p. 61).

The MIT researchers seemingly operate under the assumption that it is morally imperative to adhere to the desires of the majority to program the software of AVs. For the purpose of distinction, I will refer to this type of AVs as preference-satisfaction AVs (PAVs), in contrast to hedonistic UAVs, which are programmed to maximize objectively-calculated overall pleasures. While PAVs differ from UAVs in that they are no longer strictly hedonistic, they still fall within the realm of consequentialism. PAVs evaluate consequences not in terms of maximizing net pleasures but by satisfying a greater number of individual preferences than alternatives. They are not solely SAVs either, since they consider a wide range of individual preferences, including those that pertain to other people, in addition to their own self-interests, to make "ethical" decisions. In addition, they adhere to the market principle by allowing customers to determine the priority ranking of their AVs, thereby promoting effective marketing strategies. Indeed, PAVs have seemed to be the most successful approach to AV ethics.

A Deontological Approach

Unlike consequentialism, which evaluates decisions or actions solely based on the outcomes they produce, such as overall pleasures or satisfaction of preferences, deontological theories assess decisions or actions in terms of their inherent qualities that make them morally right or wrong. In deontology, certain decisions or actions are considered inherently evil and should never be morally permitted, regardless of any potential benefits they may bring about. In response to the consequentialist approach adopted by the MIT platform mentioned earlier, John Harris has presented a notable deontological approach to AV programming (Harris, 2020). While John Harris is not commonly regarded as a deontologist, his essay on AV ethics clearly demonstrates a deontological perspective, distinct from consequentialism or virtue ethics, which warrants careful evaluation.[3]

In his essay, Harris launches a vehement critique of MIT's consequentialist morality machine. According to him, the approach of determining

[3] While some scholars have categorized Harris as a "libertarian consequentialist" (Watts, 2007), this classification does not diminish the deontological nature of the argument he presents for AV ethics under discussion. Given its distinction from other deontological versions like Asimov's three laws (Anderson, 2008), Kantian theory (Powers, 2006), or the Rawlsian difference principle (Leben, 2017), Harris' viewpoint holds greater relevance for the purpose of this paper.

"public morality" by relying on majority preferences for programming AVs, as adopted by the MIT approach, is highly misleading. Harris argues that majorities cannot be automatically considered correct, just as neither science nor ethics is determined by voting for correct "answers" (Harris, 2020, p. 74). Instead, Harris emphasizes the importance of considering the potential harmful effects of preferences in the decision-making process. Factors such as due process, the principle of giving every person a fair hearing, constitutional protections for individual freedom, and justice, including criminal justice, should all be taken into account when establishing public morality for AV software programming (p. 75). Harris suggests that laws and the deliberative process of lawmaking serve as indicators of public morality. Preferences alone require the involvement of parliaments, legislators, courts, human rights conventions, and other stakeholders who engage in debates, consultations, and legislative actions to address and accommodate them (p. 75). Harris argues that MIT's consequentialist-programmed PAVs, like self-protective vehicles (SAVs) or any approach that condones sacrificing those outside the AVs to save the occupants inside, are morally indefensible (p. 76).

Then how should AVs be programmed in a morally defensible manner? According to Harris, the following deontological principle should be adhered to: no one should intentionally cause the death of an innocent person unless there is an overwhelming, plausible, and judicially approved necessity (p. 78). He argues that all bystanders should be presumed innocent until proven guilty through due process, regardless of whether they are elderly, breaking the law, jaywalking, or simply being lazy (p. 78). He emphasizes that humans, as moral agents, should follow a moral principle that requires self-sacrifice to save the lives of their fellow humans. He cites the example of British adventurer Oates, who demonstrated this moral principle during the Antarctic expedition in 1911. Oates left the tent and walked to his death to prevent burdening his colleagues and causing their deaths (p. 77). This "Oates Law," for Harris, should serve as the appropriate moral principle to guide the programming of AVs when faced with ethical dilemmas.

Specifically, Harris suggests that in ethical dilemmas, AVs should be preprogrammed to prioritize steering towards a solid barrier rather than selecting any human targets (p. 75). Let's refer to the AVs programmed in this manner as deontological AVs (DAVs). DAVs clearly differ from SAVs, which aim to maximize the interests of their users, UAVs, which seek to maximize overall happiness, or PAVs, which aim to satisfy the preferences of the majority. DAVs are designed based on a strict deontological principle that emphasizes the self-sacrifice of human agents to save the lives of their fellow humans.

It is worth noting that although Harris' argument is not consequentialist, he does partially consider the consequences to justify his conclusion. As he sees it, since AVs are equipped with collision technology, crumple zones, and airbags to protect the occupants, they are less likely to be fatally injured when steered towards a solid barrier. Nevertheless, Harris' argument is primarily deontological because it is based on a non-consequentialist principle focusing on a moral intention and proper decision-making procedures. Importantly, when addressing the potential lack of popularity of DAVs due to these priorities, Harris comments: "perhaps that is for the best; until … we have truly autonomous vehicles that can take on moral responsibilities seriously and have received proper education, including in law and ethics" (p. 78). This implies that Harris prefers to stick with the current scenario of human-driven vehicles while embracing DAVs, despite the fact that the outcomes are not as favorable as using SAVs or PAVs.

A Virtue-Ethics Approach

Instead of focusing on consequences (consequentialism) or principles (deontology), virtue ethics places emphasis on cultivating good character. According to virtue ethics, a virtuous person is capable of making the right decisions at the right time and in the right manner. In their notable book on machine ethics, Wallach and Allen (2009) propose an Aristotelian virtue-ethics framework for developing artificial moral machines, including AVs. The authors argue that virtues cannot be neatly categorized as either top-down or bottom-up approaches; rather, they are hybrids (p. 120). The authors advocate for a complementary relationship between the bottom-up forces that shape an individual's moral sensibility and the top-down considerations concerning the individual's relationship with the community in developing moral machines (p. 123).

Wallace and Allen contend that human moral decision making lacks a formal procedure. Instead, it is a complex process that includes emotions, intuitions, automated responses in the form of heuristics, rules and duties, and potentially explicit assessments of utility or expected outcomes (p. 177). Therefore, they endorse an approach to machine ethics based on the "learning intelligent distribution agent" (LIDA), a conceptual and computational model of cognition developed by computer scientists and neuroscientists (p. 172). Such a LIDA-based machine ethics accommodates the messiness of human ethical decision making and provides a useful framework to mechanistically represent the various influences on ethical decisions, such as feelings and rules (p. 179). In LIDA, scenarios are built and evaluated by thousands of different individual codelets performing discrete tasks (p. 183). Adopting this methodology, the authors believe that developing ethical machines through a virtue-ethics approach will yield fruitful results (p. 110).

A Hybrid Approach

Some view that each of the theoretical approaches mentioned above is limited and incomplete when it comes to designing ethical AVs because they focus solely on individual ethical components. Such limitations, according to this viewpoint, have resulted in centuries of debate regarding which moral principles or theories are most suitable for guiding human moral evaluation. As a result, some researchers have turned to theoretically integrated approaches to AV ethics, with the agent-deed-consequence (ADC) model gaining prominence. The ADC model proposes that moral judgment involves simultaneous assessments of three intuitive components: the character of the agent (Agent-component, A), their actions (Deed-component, D), and the resulting consequences in a given situation (Consequences-component, C). By doing so, the model explains the intuitive appeal of ideas derived from three dominant moral theories: virtue ethics, deontology, and consequentialism. Proponents of this model believe that it best captures the essence of human moral judgment, which is both adaptable and stable. It suggests that moral judgments are made rapidly through heuristic-like intuitive processes, without requiring deliberate effort (Dubljević et al., 2018).

These intuitive assessments combine to form an overall judgment of moral acceptability, either positive or negative. For instance, when evaluating a situation where someone has committed a negative deed, such as breaking a rule, the judgment of moral acceptability can be moderated if the agent had good intentions and the deed resulted in positive consequences. This phenomenon elucidates the significant flexibility and consistency observed in human moral judgment, which can be replicated in AVs (Dubljević, 2020). In other words, an incorrect action can be deemed more acceptable if both the character of the individual (the agent) and the consequences of their actions are positive. The ADC model provides formulas to represent such scenarios (e.g., [A+], [D-], and [C+] = [MJ+] or positive moral judgment). Advocates of this model assert that it has been empirically validated in a recent study that incorporated intuitive aspects of all three moral theories, involving 152 professional philosophers and 1314 individuals without ethics training (Dubljević et al., 2018).

When adopting the ADC model, proponents argue that programming AVs does not involve explicitly encoding duties or intentions. Instead, AVs are equipped with functional equivalents or even "limited functional simulacra" of morality. For instance, deontological rules are already integrated as AVs possess a functional equivalent of "observing the law" (the D-component in the ADC algorithm). To explicitly minimize harm in collision situations (the C-component in the ADC algorithm), avoidance algorithms utilizing facial recognition technology can be employed to limit the number of humans at risk. Additionally, the identification of malicious actors can be accomplished through the utilization of "Identification Friend or Foe" (IFF) technology. To further enable functional equivalents of moral decisions in the hybrid environment where AVs coexist with humans, an additional designation of "malfunctioning" could be incorporated into the system (Dubljević, 2020, pp. 2467–2468). In short, proponents argue that incorporating this model into the programming of AVs would be most appropriate and ethically better than any non-hybrid approach.

Towards a Culture-Relevant Approach

Confucianism as the Mainstream Moral Culture of East Asian Regions

Each of the aforementioned approaches is maximal universalist approaches. Their aim is to establish uniform ethical codes for determining the algorithms of AVs everywhere, irrespective of cultural variations across regions or societies. However, such ethical frameworks are only consistent with the criteria of some moral cultures, but not others, no matter how such cultures are identified in theoretical or ideological format. Hence, such approaches lack justification, since there is no means to assert the superiority of any of them detached from a particular cultural context. For East Asian regions, none of them align with their prevailing moral culture, notably Confucianism.[4] Although it is widely agreed that Confucianism presents the form of virtue ethics rather than consequentialism or deontology, it adheres to a distinct version of virtue ethics, diverging from the Aristotelian conception. While Aristotelianism is the predominant version of virtue ethics in the West, Confucianism constitutes its counterpart in the East, encompassing unique intricacies and subtleties in understanding the moral virtues manifested in the people's ritual practices (Fan, 2010).

Indeed, compelling evidence indicates that many East Asian regions, such as mainland China and Hong Kong, continue to adhere to Confucian moral culture. Despite the presence of various religions in these societies and their official political ideologies no longer being Confucian, individuals' ethical behaviors remain primarily reflecting Confucian morality. Their typical language usage is characterized by an extensive vocabulary of Confucian virtues, such as humaneness (*ren*仁), filial piety (*xiao* 孝), righteousness (*yi* 義), wisdom (*zhi*智), harmony (*he* 和), courage (*yong* 勇), sincerity (*cheng*誠), fidelity (*xin* 信), loyalty (*zhong*忠), and reciprocity (*shu*恕) (Fan, 2021).[5] Cultural psychology studies have provided robust evidence, obtained through rigorous experimentation, demonstrating the enduring influence of Confucian values on these regions (Li, 2012; Nisbett, 2003). In recent years, mainland China has witnessed a resurgence of Confucianism, not only as a political philosophy but also as a part of social activities and personal experiences, embracing the ideal of a

[4] Some may challenge my perspective on identifying Confucianism as the mainstream moral culture in many East Asian regions. They may argue that in each of these regions, there are individuals who believe in and practice cultures other than Confucianism, such as Buddhism and Daoism. They may contend that claiming Confucianism as the prevailing culture in these regions is misleading, and it would be inappropriate to recommend programming AVs based on Confucian ethics. They may believe that by doing so, those who reside in these regions but do not share Confucian values would be deprived of their freedom. In response to this viewpoint, I acknowledge that not every region in an East Asian country adheres to Confucian moral culture. Tibet, for example, embraces Vajrayana Buddhism rather than Confucianism. It is essential to respect their freedom to follow their predominant Buddhist culture in designing their AVs. However, each region should have the right to program AVs in alignment with their mainstream moral culture when there is no objective way to determine which culture or theory possesses the moral truth without assuming the answer. This should be done as long as each region upholds the fundamental rights and liberties of every individual in their region. It is important to note that the freedom to program AVs according to one's own values is not included in the list of basic individual rights and liberties. I express gratitude to JeeLoo Liu for engaging in discussions on these matters.

[5] All the translations of the Confucian concepts and classics used in this chapter are my own, with references to D. C. Lau (1970, 1979) and James Legge (1971).

Confucian way of life in practice (Fan, 2011; Billioud & Thoraval, 2015). From a sociological standpoint, Confucian activists have emphasized the social significance of ritual performance in Confucian temples in mainland China, thereby reinforcing Confucian ethical values (Sen, 2013). In summary, although I do not have enough room to offer a comprehensive defense here, it is reasonable to assert that it is Confucian virtue ethics, rather than any other morality, that constitutes the mainstream moral culture in numerous East Asian regions.[6]

This section will not provide a complete picture of Confucian virtue ethics. That would require a systematic explication of the Confucian good life and human flourishing, cardinal Confucian virtues, and the ways of virtue cultivation in Confucian tradition (Van Norden, 2003). In the remainder of this section, I will first outline a few key features of Confucian virtue ethics and their implications for contemporary machine ethics, then elucidate the reasons why each of the aforementioned culture-irrelevant AV approaches poses challenges for societies influenced by Confucian virtue ethical convictions, demonstrating their incompatibility. I will reserve the discussion of the Confucian perspective on determining the programming of AVs for handling brake failure circumstances in the next section.

The Confucian virtuous way of life is centered around a "human-oriented" or "putting humans first" (以人為本) approach. It advocates for social policies, organizational management, and technological advancements that prioritize the well-being, needs, and interests of human beings. In Confucian view, humans possess certain qualities, such as compassion for others' suffering, a sense of shame, a recognition of propriety, and a discernment of right and wrong, to be able to cultivate and develop moral virtues, such as humaneness (*ren* 仁), righteousness (*yi* 義), filial piety (*xiao* 孝), and loyalty (*zhong* 忠). Although animals also possess certain qualities, yet they are significantly inferior and incomparable to those of humans. Following the Confucian moral way, it is reasonable to hold that to create machines to possess superior human qualities, even if technological feasible, is morally misguided. It is of utmost importance that we should not shift the priority from "putting humans first" to "putting machines first" to advance machine welfare. Confucian machine ethics must reject the feasibility of imbuing machines with potential moral virtues.

[6] I would like to express my gratitude to Daniel A. Bell for engaging in a valuable discussion with me on this topic and for providing recommendations on relevant literature.

In contrast to Aristotle's perspective, which regarded humans as inherently political animals involved in public decision making, Confucius emphasized the essential nature of humans as familial beings. He emphasized the significance of cultivating appropriate familial relationships and extending the love inherent in those relationships to all individuals through ritual practices (as elaborated below). At the core of the fundamental Confucian virtue of humaneness (*ren* 仁) is love (*ai*愛) (Analects 12.22). Importantly, Confucian love encompasses both universal and differentiated aspects. On one hand, individuals are encouraged to love all human beings because they are considered to be the most noble (*gui* 貴) in the world (Analects 1.6; Mencius 7A:46). On the other hand, it is natural and appropriate for individuals to have a deeper love for their family members than others (Analects 13:18, Mencius 3A:5, 7A:5). It is also believed that by practicing familial love, especially giving special care to the young and the old within the family, one is able to extend such love to other people, shaping a humane ethos of caring for the young and the old in whole society (Analects 5:26, Mencius 1A:7). Thus, when Confucius advocates leaning upon humaneness for support in life, it can be reasonably interpreted as prioritizing the family within that context. Among its numerous implications for machine ethics, it implies that society should refrain from creating or utilizing machines in ways that restrict or deny individuals the opportunity of exercising graded and differentiated love in favor of their families. For instance, if individuals were compelled to employ a UAV that had been designed to sacrifice their family members inside the vehicle for the sake of promoting general happiness, it would contradict the value of Confucian family prioritization.[7]

Finally, Confucianism acknowledges that virtues must be pursued and demonstrated through the engagement with learning and observing ritual practices (*li* 禮). This is why Confucian behavior is primarily guided by

[7] Of course, Confucians face challenging situations when balancing universal and differentiated love. For instance, consider a scenario where you encounter a brake failure. According to Confucian virtue ethics, it may be deemed appropriate to prioritize the safety of your three family members inside the vehicle by choosing to stay on course rather than turning towards a solid barrier, even if it means endangering the three bystanders outside. This preference is justified based on the principled reasons derived from Confucian ethics. However, what if the number of outside bystanders increases to four or five instead of three? In such cases, I believe it would be inappropriate for Confucianism, as a virtue ethics framework, to provide a universal answer without considering the specific context. I will explore this issue further in Section "Confucian Justifications".

ritual practices. Confucian ritual practices encompass a range of familial and social behavioral patterns and conventions, encompassing both ceremonial rites (such as those associated with coming of age, marriage, funerals, and ancestral sacrifices) and everyday gestures (such as smiling, nodding, and bowing), which provide guidance for individual actions. It is important to emphasize that these practices are not established through a top-down deductive approach to universal principles or rules. Instead, they have evolved organically within Chinese society, driven by local customs and traditions. Furthermore, these rituals are not imposed regulations enforced by the government to compel compliance. Rather, it is the personal choice of each individual and family to decide whether to observe a particular ritual in a given context. Indeed, "ritual" and "law" are fundamentally distinct. Rituals are not coercively enforced by the government but rather left to individuals to cultivate oneself and restore ritual (Analects 12.1), involving thoughtful consideration and reflection. In this way, rituals effectively guide individuals' voluntary choices in speech and conduct.[8]

In its structure, Confucian ritual practices are shaped and guided by a system of constitutive rules and regulative principles. These rules and principles are inherent within the rituals themselves and are also formulated to instruct new learners and practitioners. A constitutive rule takes the form of "in context c, doing x constitutes as y." On the other hand, a regulative principle provides overarching guidance in the format of "do x" or "do not do y." For instance, in the *Analects*, Confucius articulates a series of regulative principles to direct individuals in performing ritual practices appropriately. He often explains the virtue of humaneness (*ren*) in terms of more specific virtues, which can be seen as a set of specific Confucian regulative principles. For example, he states, "while at home, hold yourself in a respectful attitude (*gong*恭); when serving in an official capacity, be reverent (*jing* 敬); when dealing with others, be loyal (*zhong*忠)" (Analects 13.19). Additionally, Confucius proposes general regulative principles, including the famous golden rule: "do not impose on others what you yourself do not desire" (己所不欲，勿施於人) (Analects 12.2).

Constitutive rules provide guidance on "what to do" (such as offering specific food and drink to ancestors on a memorial day, which constitutes as remembering them), while regulative principles instruct on "how to do" (such as maintaining an attitude of seriousness and reverence during

[8] For a detailed account of the Confucian rituals, see Fan (2010, Ch 11).

a memorial ritual). It is crucial to recognize that regulative principles cannot and should not replace constitutive rules in guiding life activities. Different interpretations of a regulative principle can lead to disagreements about its specific requirements in certain cases. Constitutive rules and regulative principles must work together to play a comprehensive role in guiding human conduct. There may be situations where they come into conflict with each other. When such conflicts arise, there is no absolute formula to determine which should override the other. Instead, one needs to engaged in a reflective equilibrium between them and rely on practical wisdom (*zhi* 智) to resolve the issue.[9]

Indeed, Confucius generally recommends "observing the rituals" as a means to guide individuals in carrying out righteous actions. This recommendation is sensible because individuals are often swayed by their self-centered desires, leading them to neglect the proper observance of ritual practices (Analects 12.1). However, relying solely on established ritual practices is insufficient to ensure adherence to Confucian ethics. Firstly, there are ritual practices that are flawed and should be abolished. Additionally, certain ritual practices may require adaptation or reform in response to changes in social or economic conditions. Furthermore, there are situations where exceptions to observing a ritual practice may arise in specific contexts. Lastly, as humans encounter new situations that have not been addressed by traditional ritual practices, they must develop new practices to suitably adapt. For instance, in the context of driving automobiles, various ritual practices have emerged to promote good driving etiquette (in addition to adhering to traffic laws). Examples include yielding to pedestrians and taking turns at intersections when no traffic lights are present. However, there may not be any established ritual practice to guide us on how to react to an emergent and complex dilemma. In such situations, individuals must rely on practical wisdom to determine the appropriate course of action within the given context. Indeed, when it comes to programming AVs to handle brake failure scenarios, there are no existing ritual practices for us to follow in thinking about what suggestions we should make.

[9] For a detailed Confucian account of the relationship between constitutive rules and regulative principles, see Fan (2024).

The Problems of Culture-Irrelevant Approaches in East Asian Regions

As a type of virtue ethics, Confucianism inherently cannot endorse the ethical legitimacy of selfish AVs (SAVs) that prioritize the interests of occupants exclusively. Such designs fail to align with the virtue requirement of caring for others, not just oneself. Similarly, Confucianism cannot affirm the ethical legitimacy of utility-focused AVs (UAVs), because they solely seek to maximize general pleasures. Virtue concerns extend beyond hedonistic consequences and care about proper moral character and behaviour motives. This raises the question of how Confucianism views preference-satisfaction AVs (PAVs), as proposed by the MIT project.

As discussed in section "Culture-Irrelevant Approaches", the surveys conducted in Western societies indicate a contradictory situation where individuals acknowledge the ethical superiority of UAVs while expecting others to adopt them, yet personally opting for SAVs to prioritize their own safety. Although there are cultural distinctions between Confucian Eastern and liberal Western societies, it is unfounded to assume that East Asian regions exempt from this discrepancy. Consequently, some might contend that just as preference-satisfaction-based PAVs can address the concerns of both SAVs and UAVs by incorporating the preferences of the majority in Western societies, they would likely be more widely accepted by most East Asian individuals compared to UAVs or SAVs. This implies that people in East Asian regions would be more inclined to purchase PAVs than UAVs. They might conclude: given that majority preferences are significantly influenced by cultural factors that vary across different regions, and considering the prominence of Confucianism as the prevailing moral culture in East Asia, we should rely on the majority preferences discovered through surveys made in relevant East Asian regions to design AVs for these regions. In other words, PAVs for these regions would undoubtedly reflect the essence of Confucian morality.

They might offer the following example to illustrate their conclusion. Influenced by the Confucian virtue of filial piety, which emphasizes respect and care for the elderly, a significant number of individuals in East Asia would disapprove of programming AVs to prioritize saving young lives over the lives of the elderly in emergency situations. Instead, they may even prefer to sacrifice young for the lives of the elderly. If this is indeed the finding through a survey, they would emphasize, then designing AVs as PAVs by following the majority preferences would definitely align with Confucian ethics and be suitable for East Asia to adopt.

However, I find this argument to be unsound for several reasons. First, Confucianism does not assert that the right morality can be determined solely through public opinions or majority preferences. It is widely recognized that certain preferences are problematic in themselves and may even possess morally objectionable content. For instance, Ronald Dworkin (1977, pp. 234–237) highlights discriminatory preferences as having an "immoral content" that should not be considered when making morally just decisions. Additionally, preferences influenced by "pathological processes," such as the minimal preferences of a disabled individual discussed by Amartya Sen (1985), should not be taken literally when formulating public policies. In the context of designing AVs and seeking majority preferences through public surveys in East Asia, it is unclear how one can effectively exclude the influence of "immoral" or "pathological" preferences from the resulting majority preference. Therefore, relying solely on such a majority preference may not genuinely reflect the genuine ethical convictions of Confucian virtue ethics.

The teachings of Confucius do not support the notion that Confucian moral judgments should be based solely on a majority or unanimous opinion. In one instance, when a student asked Confucius if they should approve of someone when everyone in the village loves them, he responded, "that is not enough." Similarly, when asked if they should consider someone bad when everyone in the village hates them, he replied, "it is not enough either." His view is that "it would be better if those who are good in his village also like him and those who are bad dislike him" (Analects 13.24). The essence of Confucius's teaching is that Confucian virtue ethics has its own set of standards for virtue that cannot be determined solely by public opinions. Although public opinions are one factor for reference, other considerations must also be taken into account. In the following section, I will explore these additional standards in more detail. For now, it is sufficient to note that this perspective demonstrates that Confucian society cannot endorse the legitimacy of PAVs.

On the other hand, although the proposal of Harris' DAVs, which are programmed to always swerve towards a solid barrier in a crisis situation to protect outside bystanders, may appear morally altruistic and admirable to some individuals in East Asia, I believe they are also not morally consistent with Confucian virtue ethics when it comes to designing AVs. My reasoning primarily revolves around two key points.

First, these DAVs are ethically inadequate as they fail to consider the relevant states of affairs in the deliberation process of programming AVs. While Confucian moral culture prioritizes the practice of virtues over improving consequences, it is important to note that it does not, and should not disregard consequence considerations entirely. As previously mentioned, if the adoption of AVs leads to a significant reduction in casualties and injuries compared to the current scenario with human-driven vehicles, the value of such better consequences should not be ignored by Confucian society. Harris appears to disregard these different states of affairs entirely, focusing solely on justifying his DAVs based on a deontological principle that prohibits deliberately killing innocents, even if it means maintaining a significantly worse state of affairs. Adhering to this deontological principle, Harris is willing to accept more injuries and deaths caused by human-driven vehicles without making any attempt to improve the situation through the introduction of AVs, until they will eventually become morally responsible agents themselves. In this regard, Confucian virtue ethics would side with the usual critique of deontology from consequentialists, who argue that by preserving the purity of moral agency, deontological decisions may lead to a morally worse state of affairs. Such an approach cannot be embraced by Confucian virtue ethics, although it is not merely consequentialist.

Furthermore, Harris overlooks the fact that AVs will inevitably be utilized for various purposes. For instance, automated trucks will primarily be employed for transporting material goods or animals rather than human passengers. It is logical to programme such trucks to always swerve onto a solid barrier in crisis situations to protect human bystanders. This aligns with the strong Confucian moral belief that human lives hold greater importance than material possessions or animals. However, it becomes difficult to comprehend why buses, such as a school bus carrying dozens of primary school students, should also be programmed to swerve onto a solid barrier, risking severe loss of life, to avoid injuring one or two bystanders. In this context, maintaining the bus's course would result in less harm, consistent with Confucian virtue concerns.

In short, Harris' deontological approach lacks sensitivity to the diverse contexts in which AVs are used. In a society influenced by Confucian virtue ethics, which emphasizes virtuous human relationships with one another and with the world, policy formulation along with the establishment of ethical codes for AVs must account for these different contexts. Instead of applying a single absolute deontological principle to every situation, we need to consider programming AV software in a way that allows individuals to utilize AVs to facilitate their virtue cultivation and the formation and maintenance of virtuous relationships with others. To integrate all these considerations, Confucian AVs should be programmed differently from both consequentialist and deontological AVs.

Wallace and Allen's Aristotelian virtue-ethics framework, which combines top-down and bottom-up strategies, has garnered significant recognition and acclaim. Their approach focuses on specific Aristotelian virtues such as courage, justice, and temperance. However, the question arises as to how these virtues can be translated into tangible and measurable attributes for machines. Moreover, even if such translation is possible, it is reasonable to raise concerns regarding the compatibility of these virtues with the Confucian virtues prevalent in East Asian societies, such as humaneness (*ren*), righteousness (*yi*), and filial piety (*xiao*). In response to this challenge, East Asians would reasonably argue that machines used in East Asia should be programmed with Confucian virtues rather than Aristotelian virtues. These societies perceive humaneness, righteousness, and filial piety as genuine virtues that hold significant cultural and moral value. The commensurability of these virtues with the Aristotelian virtues remains uncertain.

Importantly, Wallace and Allen's approach aims to develop artificial autonomous moral agents capable of engaging in ethically significant actions alongside human individuals. Regardless of the technical feasibility of this virtue-based strategy, Confucian virtue ethics may not endorse such an endeavor. While it is conceivable for humans to create software that *implicitly* supports ethical behavior, such as automated teller machines and web banking software designed to accurately process financial transactions and uphold the ethical value of honesty, it is an entirely different matter to create *explicit and fully-fledged* artificial ethical agents. These agents would autonomously navigate real-life situations involving unpredictable sequences of events and even possess consciousness, intentionality, and freedom in both performing and justifying their actions (Moor, 1979, 2006). In the former scenario of "implicit" artificial agents, humans make ethical decisions and utilize machines in an ethically responsible manner. However, in the latter scenario of "explicit" artificial agents, humans would be responsible for creating ethical machines that are explicit or complete moral agents, capable of independently making ethical decisions. This distinction highlights the significant ethical implications and challenges associated with developing machines that possess the capacity for moral agency.

The extent to which artificial agents can achieve certain capabilities remains a subject of empirical investigation. However, the development of such agents cannot be supported by Confucian virtue ethics, primarily due to its emphasis on human life and human relationships.[10] According to Confucian morality, humans have a responsibility to procreate and educate their offspring, guiding them towards acquiring virtues through ritual practices. This process necessitates allowing individuals to make moral mistakes as part of their learning and development, ultimately becoming fully mature moral agents. Even when individuals reach this stage, they retain the freedom to act contrary to virtue, as it is through these choices that they can genuinely exercise virtuous behaviour and foster proper human relationships. This dynamic of individual freedom is integral to the Confucian virtuous way of life.

However, there is no compelling reason for Confucians to create machines with the capacity to make moral mistakes to alternate between ethical and unethical behavior. Confucians believe that humans hold the highest value among all creatures and should naturally possess this freedom, maintaining a relationship of mutual respect with one another. Machines, on the other hand, are designed by humans as tools to enhance human wellbeing. It is illogical and contradictory for humans to grant machines such freedom, especially considering the potential disasters that might arise from their exercise of such freedom. In short, Confucian moral culture does not find Wallace and Allen's virtue-ethics approach appropriate, as it contradicts the foundational conception of Confucianism related to the insurmountable value of human life and relationships.

The ADC model is often regarded as more pluralistic compared to other approaches. However, it may not be suitable for the Confucian culture of East Asia. This model aims to integrate the character of the agent (Agent-component, A), their actions (Deed-component, D), and the resulting consequences in a given situation (Consequences-component,

[10] My critique of Wallace and Allen's approach from a Confucian standpoint is influenced by R. Token's observation of a "troublesome dilemma" brought to light by their approach: on one hand, virtue ethics appears to welcome the creation of autonomous and virtuous machines; on the other hand, the risk of such machines acting immorally is necessary for them to become virtuous (Tokens, 2012, p. 144). However, Token's analysis does not specifically explore the issue through the lens of Confucian virtue ethics focusing on the value of human life and relationships.

C) to establish a comprehensive moral code for societies to adopt. However, it faces challenges related to unclear weighting or prioritization among its different components. If the model simply suggests that two positive components outweigh one negative component, it can lead to inconsistent or arbitrary outcomes in practical applications.

Furthermore, the ADC model lacks cultural and contextual sensitivity. The intricacies lie in the details. What defines a good character or intention? How do we determine a right action? And how should consequences be measured? Without considering the nuances of Confucian culture, it becomes impossible to provide specific guidance for the model in East Asian regions. Finally, despite encompassing three broad components, the model still has limited scope in terms of moral considerations. It may overlook significant Confucian ethical concerns such as the inherent value of human life, the importance of appropriate and harmonious human relationships, the concept of universal but differentiated love, and the significance of filial piety, among others. These values hold great importance in East Asian societies and cannot be adequately addressed by this general hybrid model for designing AVs.[11]

Confucian AVs

Confucian Methodological Considerations

How can Confucian virtue ethics be appropriately applied to guide the programming of AVs in Confucian-influenced regions when faced with brake failure scenarios? In section "Confucianism as the Mainstream Moral Culture of East Asian Regions", it is noted that even for human-driven vehicles, there are no established ritual practices to teach drivers how to handle such complex situations, although there are certain rituals available to guide regular driving activities. Indeed, the Confucian constitutive rules and regulative principles embedded in the ritual practices provide guidance for individuals to navigate everyday activities, develop good habits, and cultivate virtues. However, in emergency or exceptional dilemmatic cases, individuals must rely on practical wisdom to determine the right course of action. Therefore, Confucian virtue ethics is ultimately

[11] Similar challenges would likely arise with other popular hybrid or integrative ethical approaches to AVs, such as the framework proposed by Rhim et al. (2021).

associated with practical wisdom in addressing difficult situations, and practical wisdom cannot be fully encapsulated by rigid principles.

Some may argue that Confucians should adopt the Daoist perspective of non-action (*wuwei* 無為) and avoid establishing ritual practices even for ordinary activities. According to this view, Chinese Daoists, particularly Zhuangzi, believe that offering rules or principles to guide action is unnecessary. They contend that action should arise spontaneously and be guided by practical wisdom. Instead of prescribing rules or principles, a wise person reduces wisdom to its essence, which is "the dispassionate mirroring of things as they are" (Graham, 1991, pp. 191–194). However, I believe that Confucianism would uphold a different stance. It would assert that the guidance of constitutive rules and regulative principles is essential, especially for inexperienced individuals, as it aids their learning process and helps them eventually attain practical wisdom. In other words, rules and principles are at least beneficial in directing young individuals in handling ordinary cases, if not more complex ones. Indeed, Confucius consistently emphasized the significance of learning ritual practices and acknowledged that he did not fully achieve the state of perfect wisdom, where desires align with what is morally right, until he reached the age of seventy (Analects 2.4).

What about highly challenging cases? Some argue that Confucians should refrain from providing specific guidelines for programming AVs to handle brake failure scenarios, as these cases are unique and dilemmatic. Similar to Aristotelian virtue ethics, Confucian virtue ethics emphasizes the importance of practical wisdom in dealing with difficult or dilemmatic situations. Practical wisdom cannot be acquired solely through book learning, as one would learn and apply mathematical or logical formulas. Instead, individuals must cultivate practical wisdom through experiential activities and real-life encounters. They must engage in ritual practices where they determine when to deviate from established patterns, reform them, or create new ones, with the aim of becoming practically wise. Therefore, for them, Confucian ethics should not aim to determine the digital algorithms of AVs for handling brake failure situations.

This view aligns with the strategy employed by a comprehensive German ethics code for AVs, published in 2017. According to the code, Ethical Guideline 8 states that "genuine dilemmatic decisions, such as a decision between one human life and another, depend on the actual situation, incorporating 'unpredictable' behavior by parties affected. They can thus not be clearly standardized, nor can they be programmed such that they are ethically unquestionable. ... [such decisions] cannot readily be transformed into abstract/general *ex ante* appraisals and thus also not into corresponding programming activities" (Luetge, 2017, p. 552). However, while it is true that such decisions cannot be fully programmed into digital algorithms to handle complex cases, it is incorrect to suggest that nothing relevant should be done in programming AVs. In fact, when ethics refrains from offering any guidance on programming for handling challenging cases, engineers are left to conduct the programming work according to their own discretion. Typically, they would program an AV to maintain its course in a dilemmatic situation and apply maximum brake force, even if they "know" that the distance is insufficient for the AV to stop before colliding with a human individual. This approach does not avoid making an ethical decision; rather, it involves the decision to prioritize one ethical principle in the AVs' algorithms: protecting the occupants of the AVs by staying on course. From a Confucian virtue ethical perspective, this decision may be deemed appropriate in certain cases but inappropriate in others.

This implies that Confucian virtue ethics should not abstain from providing necessary ethical guidance for such cases. However, it is important to note that developing a Confucian ethical code should not be approached in the same manner as utilitarianism or deontological ethics, which aim to establish rigid guidelines for determining correct individual actions or policy decisions in all circumstances. In other words, attempting to create a Confucian ethical strategy for designing Confucian AVs, akin to how utilitarians offer guidance for UAVs or Harris' deontological approach for DAVs, with a single standardized criterion (either always maximizing utility or always sacrificing occupants) to address all challenging cases, would not be appropriate. Instead, as discussed in section "Confucianism as the Mainstream Moral Culture of East Asian Regions", Confucian virtue ethics relies on a reflective equilibrium between constitutive rules and regulative principles to navigate complex issues. While Confucian virtue ethics has formulated a set of regulative principles to provide overarching guidance, it is crucial not to consider any specific regulative principle or a particular interpretation of it as always dominant when programming AVs to handle brake failure scenarios.[12] This approach distorts the essence of

[12] JeeLoo Liu, in her thought-provoking and innovative paper (Liu, 2021), constructs a distinct Confucian regulative principle that she considers to be central in her formulation of Confucian AV ethics. This principle is built upon her unique interpretation of the Confucian

Confucian virtue ethics, which relies on a reflective equilibrium, and undermines the complexity of challenging cases.

Confucian virtue ethics acknowledges that each problem possesses unique contextual features, and it is crucial to take these features into account when seeking appropriate solutions. When addressing the issue of programming AVs to handle potential brake failure scenarios in East Asian regions, there are notable contextual considerations that must be addressed. One such consideration revolves around the importance of nurturing virtues and upholding family-oriented concerns. As discussed in section "Confucianism as the Mainstream Moral Culture of East Asian Regions", the use of AVs should not undermine individuals' ability to express differentiated and graded love towards their family members in reasonable ways. Another significant contextual feature is the presence of

virtue of loyalty (*zhong* 忠) and is intended to take precedence over other regulative principles. Liu's perspective posits that a Confucian AV's primary obligation is to fulfill its designated role of safeguarding its occupants. Accordingly, its sole "duty" is to ensure the occupants' safety by never taking actions that could result in colliding with a barrier or causing a catastrophic disaster. Liu's understanding of the Confucian virtue of loyalty characterizes it as a relationship-oriented virtue directed towards one's assigned role. She defines Confucian loyalty as "doing what one is supposed to do" or "being loyal to one's role." Consequently, she designates "loyalty to one's task" as the foremost Confucian principle in the context of Confucian AVs, asserting that AVs should be designed to prioritize their role-specific function of protecting their occupants.

While Liu's proposal may garner support from AV users, it suffers from significant flaws. Firstly, Liu fails to provide a compelling argument to demonstrate why her interpretation of loyalty is more relevant than the generally-accepted interpretation presented by Yang Bojun 楊伯峻 (1980, p. 16). According to Yang, loyalty is a relationship-oriented virtue directed towards others, as evidenced in passages such as Analects 1.4, 12.23, and 14.7. Yang's interpretation encompasses a regulative principle advocated by Confucius to guide individuals in cultivating the fundamental virtue of *ren*, which entails "helping others to establish themselves while seeking one's own establishment, and assisting others in achieving success while aspiring to one's own success" (己欲立而立人,己欲達而達人) (Analects 6.30). In contrast, Liu argues that this principle is only applicable to *ren* and not to *zhong*, without providing adequate justification. Furthermore, Liu's proposal leads to the creation of "Confucian" AVs that are essentially self-centered, as discussed earlier, which contradicts the fundamental orientation of the Confucian virtues that promote love towards all humans, albeit in differentiated and graded ways, rather than solely focusing on oneself and one's family. It is challenging to defend the notion that Confucian virtue ethics would consistently advocate for prioritizing the protection of occupants over bystanders in every case without exception. Liu excessively relies on a specific Confucian regulative principle to develop her concept of Confucian AVs without considering contextual features and engaging in a necessary reflective equilibrium.

distinct types of AVs designed for various purposes in these regions. To ensure ethical programming, it is necessary to differentiate between these vehicle types and consider their specific functions. At least three distinct types of AVs are expected to be employed. Firstly, there are automated trucks intended for transporting goods or animals, rather than human passengers. Secondly, automated buses, including school buses, city buses, and buses for elderly care facilities, are designed to transport large groups of human passengers. Lastly, automated small vehicles, such as individual or family cars, are meant to carry a single individual or a very small group of passengers. It would be beneficial to initially address each type separately when determining the appropriate programming approach from a Confucian ethical standpoint. In fact, as elaborated below, it is crucial to program them differently.

Before providing specific suggestions for programming Confucian AVs, I must address a potential methodological challenge to my approach. Some may argue that my strategy contains a contradiction. On one hand, I emphasize the reliance on practical wisdom within Confucian ethics to determine appropriate handling of powerful machines like AVs since Confucian virtue ethics cannot and should not be fully theorized or reduced to rules or principles. On the other hand, I offer specific Confucian suggestions for programming AVs, which could be seen as attempting to systematize Confucian wisdom into principles for AV programming. The challenger could conclude that if I claim my suggestions are derived from Confucian virtue ethics, then I must accept that Confucian virtue ethics can be fully reduced to principles, contradicting my original statement.

To address this challenge, I will examine the usage of the term "fully" in the original statement. As mentioned earlier, Confucian virtue ethics provides constitutive rules and regulative principles to guide ordinary situations and foster practical wisdom. While experienced individuals may not require such rules and principles, they are valuable in educating and guiding young individuals in their everyday conduct. Furthermore, when it is anticipated that a minor or student will encounter a challenging situation, it is reasonable for their parents or teachers to provide them with a rule or principle to aid their decision making, even if the absolute correctness of that rule or principle in that specific context is uncertain. This guidance is offered as long as the rule or principle is not obviously more misguided than other alternatives but proves helpful for the young person to consider and prepare for the forthcoming situation. Of course, there is no guarantee that the minor or student will unquestionably follow the guidance, as they possess a certain degree of free will (although not complete). The distinction between a minor and an AV lies in the fact that the latter lacks any capacity for free will and will unquestioningly adhere to the preprogrammed rules and principles.

From a Confucian standpoint, it would be inappropriate for parents or teachers to refrain from offering any guidance to a minor who will inevitably encounter challenging situations. Therefore, while Confucian practical wisdom cannot be fully reduced to principles to guide young individuals, it may be partly systematized to do so in exceptional circumstances. Similarly, Confucian virtue ethics cannot entirely translate its practical wisdom into the digital algorithms of AVs to handle unavoidable accident situations pre-emptively. However, it is possible to make partial attempts in this direction. This approach is preferable to designing AVs' digital algorithms for handling difficult cases without offering any Confucian ethical suggestions. Indeed, it is worth acknowledging that anything worth doing well is worth doing imperfectly.

Confucian Suggestions

Based on the aforementioned methodological considerations rooted in Confucian ethics, I would like to propose the following suggestions for programming AVs to address potential brake failure scenarios. It is crucial to note that different types of AVs should be programmed in distinct ways, taking into account their specific purposes of use.

A. For automated trucks, which are intended for transporting material goods or animals rather than human passengers, the programming should prioritize the safety of human pedestrians or bystanders. In the event of an unavoidable collision, these trucks should be programmed to swerve onto a barrier to spare any potential harm to individuals outside the vehicle.
B. Automated large-size passenger buses, including school buses, city buses, senior center buses, and other institutional buses, should be programmed to prioritize the safety of the occupants over pedestrians or bystanders in an unavoidable collision situation. Therefore, they should be designed to stay on their intended course instead of swerving onto a barrier.
C. Small AVs designed for transporting one or a few individuals (with a maximum capacity of five) should be programmed to prioritize the

safety of pedestrians or bystanders outside the vehicle in a crisis where a brake failure occurs. These AVs should be programmed to swerve onto a barrier rather than colliding with pedestrians or bystanders outside.

D. It is essential for each of these small AVs to be equipped with a steering wheel that allows a human driver to assume control at any moment. With the availability of a steering wheel, the driver can make a decision based on their own judgment to protect their family members inside the vehicle, deviating from the digital algorithm programmed as suggested in point C, if they deem it necessary in a brake failure situation.

Confucian Justifications

It is straightforward to justify suggestion A, which prioritizes the protection of human lives over material goods or animals in a crisis. The belief in the inherent and immeasurable value of human life is widely shared across different cultures. In Confucian culture, there is a firm conviction that humans hold the highest nobility (*gui* 貴) among all things in the world and should be treasured and safeguarded above all else. When programming the software for automated trucks to swerve onto a barrier to avoid colliding with a human individual, it is imperative to ensure that they are also coded in a manner that minimizes the risk of injuring any other human individuals in the vicinity of the barrier, to the extent that it is technically feasible. This additional consideration aims to prioritize human safety comprehensively while adhering to Confucian ethical concerns.

In support of suggestion B, which recommends programming large-size passenger buses to stay on course and prioritize the safety of their occupants, there are at least two Confucian considerations to consider. First and foremost, as these buses carry multiple human passengers, Confucian ethics places great emphasis on the careful protection of these lives. It is important to note that emphasizing the protection of human lives does not equate Confucian virtue ethics with utilitarian ethics, which focuses on maximizing general happiness when making ethical decisions. Rather, it highlights that considering consequences is one reasonable aspect of seeking proper Confucian ethical decisions. While consequence appeal may occasionally outweigh other considerations and play a decisive role in settling an issue, this does not reduce Confucianism to utilitarianism, as Confucianism encompasses more than just consequences in its ethical framework.

In the case of an automated bus, it is ethically reasonable and justifiable to conclude that the safety of its numerous passengers outweighs the safety of one or a few individuals on the road in an unavoidable crisis. This judgment aligns with Confucian ethics since no other relevant Confucian constitutive rules or regulative principles exist that would require a different judgment to override it. It is challenging to find any Confucian virtues that would support jeopardizing the lives of many people on a bus to spare one or a few bystanders in an accident. Therefore, prioritizing the safety of the bus occupants aligns with the Confucian virtues in this particular context.

Furthermore, it is worth noting that bus passengers often include a significant number of young children and elderly individuals, in addition to adults. Observations indicate that even on regular city buses not specifically designated for schools or senior centers, one is likely to find a higher proportion of children and elderly passengers compared to adults. This is primarily due to the fact that adults tend to prefer using their own vehicles rather than utilizing public transportation. Significantly, within Confucian virtue ethics, there is a belief that, all else being equal, society should place greater emphasis on safeguarding the lives of the young and the elderly. This emphasis stems from the recognition that these individuals are particularly vulnerable and require additional care and protection. Confucian ethics advocates for special attention and consideration towards these vulnerable groups, acknowledging their unique needs and the importance of their well-being. Therefore, prioritizing the safety of bus occupants aligns with the Confucian morality of prioritizing the protection of the young and the elderly, who constitute a significant portion of the passengers.[13]

[13] Some utilitarians may argue that it is morally right to programme a bus based on a comparison of the numbers of its occupants and outsiders in an accident. According to their viewpoint, if the number of occupants is greater than the number of outsiders, the bus should be programmed to stay on course. Conversely, if the number of outsiders is greater than the number of occupants, the bus should be designed to turn towards a barrier in order to spare more outsiders. They may advocate for this approach, even though accidents where the number of outsiders exceeds the number of occupants are extremely rare occurrences with such buses.

However, I do not support this utilitarian suggestion and instead propose a specific Confucian consideration. As mentioned in Section "Confucianism as the Mainstream Moral Culture of East Asian Regions", Confucian ethics places significant emphasis on caring for the young and the old within society. Programming buses in accordance with suggestion B

Suggestions C and D should be examined together. Confucian ethics can provide multiple reasons to support suggestion C, which proposes programming automated small cars to steer toward a solid barrier in the event of a brake failure, prioritizing the protection of bystanders. These reasons can be understood from at least three perspectives. First, within the framework of Confucian virtue ethics, it is considered highly virtuous for a person driving a human-driven car, as the sole occupant, to have the intention of preserving the lives of bystanders even at the risk of their own life by turning the car towards a solid barrier during a brake failure situation. Confucian ethics would undoubtedly commend such virtuous intentions and heroic behavior, although it does not necessarily impose such actions as a requirement. Programming small AVs according to suggestion C should be viewed as a means of promoting and encouraging such virtuous intentions, creating potential opportunities for occupants to voluntarily exhibit this behavior, without coercion. This approach is warranted because, as suggested in point D, the availability of a steering wheel allows occupants to exercise their own judgment and deviate from the programmed behavior. In light of the ethical admiration bestowed upon such intentions and behaviors within Confucian ethics, it is morally acceptable and reasonable to encourage such intentions and provide potential opportunities for their realization.[14]

serves an important ethical and educational purpose. It reinforces the notion that society should prioritize the protection of the lives of its young and old individuals due to their vulnerability and need for care. Deviating from this approach and adopting a utilitarian design would undermine the essential educational function that Confucian ethics holds in high regard.

[14] My suggestion D, which advocates for keeping a steering wheel in each small AV for a human driver to retain control, has received support from all the friends I consulted regarding this issue. However, some of my friends believe that suggestion C should be revised. According to their perspective, a small AV should be programmed to prioritize the protection of its occupants by staying on course, even if it means colliding with a human target in front. If an occupant desires to prioritize the safety of outsiders at the expense of their own, they can utilize the steering wheel to do so.

I find this suggestion to be problematic as it deviates from the fundamental concerns of Confucian virtue ethics, which emphasize caring for all individuals involved. It reduces the ethical framework to a purely selfish or egoistic idea, disregarding the ethical education and encouragement that Confucianism seeks to promote and relegating Confucian AVs to be selfish AVs. As long as programming small AVs according to my suggestion C does not involve coercion, as I believe it does not due to the presence of suggestion D, suggestion C can be better justified within Confucian virtue ethics compared to the suggestion put forth by my friends.

Moreover, the rationale behind suggestion C can be further reinforced by considering the consequences of the actions. Considering the smaller size of these cars, it is highly probable that the number of occupants inside will be fewer than the number of bystanders outside in the event of a brake failure. Moreover, these vehicles are equipped with collision technology, crumple zones, and airbags, providing a certain level of protection for their occupants during accidents. As a result, even if the car steers towards a solid barrier, the occupants may not sustain severe injuries. On the other hand, if the small AV were to collide with bystanders, they would likely suffer significant harm. Taking these consequence considerations into account adds weight to the appropriateness of suggestion C. This approach aligns with Confucian ethics by caring for the well-being of the occupants without disregarding the potential consequences for others.[15]

Finally, suggestion D, which proposes the inclusion of a steering wheel in automated small cars to allow human drivers to assume control whenever they desire, can be invoked to bolster the viability of suggestion C. These two suggestions are intricately linked, as suggestion C serves as a means to encourage and educate individuals in the exercise of virtue when operating such cars, while suggestion D grants human drivers the freedom to make choices during the process. By incorporating suggestion D, human drivers are empowered to make heroic decisions to protect bystanders even at the cost of their own lives or the lives of their family

[15] In a similar vein to the case of buses, some individuals may propose that small cars should also be programmed to make decisions based on the numbers of occupants and bystanders. However, as previously discussed in relation to buses, I believe Confucians would not support such utilitarian suggestions, as the latter prioritize the numbers of individuals involved as the sole determinant in ethical decision-making in every circumstance. Confucians would find it unreasonable, for instance, to argue that sacrificing three children or elderly passengers inside a car is always justified to spare four bystanders outside.

Moreover, if decisions were consistently made by comparing the numbers of occupants and bystanders, we would similarly be compelled to select lanes based on the numbers of bystanders. If bystanders on another lane are fewer in number than occupants, cars should be designed to turn to that lane. However, cars can also be designed to stay on the same lane and collide with a large crowd on its fringe instead of switching lanes. In certain cases, staying on the same lane may result in even fewer casualties. Such considerations and calculations would become overly complex, even without introducing additional factors like sex, age, or traffic law violations associated with the individuals involved. Adopting such a decision-making approach would necessitate the complete formalization of practical wisdom in programming AVs, which is both ethically misguided and technologically unfeasible. Taking all aspects into account, a more suitable Confucian solution would involve implementing suggestions C and D together.

members. The availability of this choice enhances the legitimacy of suggestion C, as it acknowledges the importance of individual agency and moral responsibility.

Suggestion D reflects a fundamental Confucian concern.[16] The Confucian reasons for supporting this suggestion can be summarized as follows. Firstly, qualified human drivers should be granted the option to take over AVs and control them according to their own judgment. This is because human drivers are moral agents, possessing consciousness, free will, and the ability to cultivate virtues. In contrast, AVs are artificial intelligence systems lacking consciousness and free will. Their "actions" are solely driven by algorithms programmed by humans. Therefore, it is reasonable to assert that human drivers have the right to assume control of AVs and make decisions in the given context.[17] This analogy can be compared to how humans treat animals. As mentioned earlier, Confucians believe that even if animals possess certain capacities for consciousness or freedom, their quality is significantly inferior to that of human capacities. Consequently, humans are justified in controlling and utilizing animals. Given that AVs do not possess even the capacities that animals possess, it is evident that humans are certainly justified in controlling them according to their own discretion.

However, a counterargument may arise to supplement suggestion D. Some may argue that encouraging human drivers to reject suggestion C by prioritizing their own interests or the protection of their family members contradicts the Confucian virtue ethical reasons underlying suggestion C. It is possible that some human drivers may use the availability of a steering wheel in a brake failure situation solely to consider their own well-being. While such cases cannot be entirely excluded, it is important to recognize that Confucian morality, as a virtue ethics, does not categorically assert that these actions are morally wrong in the sense of being ethically blameworthy or impermissible.

As Rosalind Hursthouse astutely points out, virtue ethics, unlike consequentialist or deontological theories, may interpret "right action"

[16] During a conversation with a Confucian friend regarding the use of small automated cars, their immediate response was, "I must have a steering wheel in my automated car to protect my family." I found this particular response to be representative of Confucian values, leading me to choose it as the chapter title.

[17] The question of whether artificial intelligence will, technologically, possess such capacities in the future remains a subject of controversy. However, it is evident that current artificial intelligent machines, including AVs, do not possess such capacities.

differently in the same situation: it can be the best action, the commendable action (though not the best possible), or the unblameworthy action (even if not commendable) (Hursthouse, 2022). For instance, it is conceivable that in a brake failure situation, according to Confucian virtue ethics, the best action for a human driver may be to turn towards a solid barrier to save a group of bystanders at the risk of one's own life. The commendable action might involve turning the car to another lane to hit one individual in order to spare the larger group. However, it may still be considered unblameworthy if the driver fails to take either of these actions and simply stays on course, colliding with the group, because they do not wish to harm themselves or the person on the other lane. While the first and second actions would be ethically superior, the last action is still ethically unblameworthy. Importantly, no one should be compelled to choose the first or second action, and no one should be penalized for taking the last action, since it aligns with the minimal Confucian ethical sense. Confucian virtue ethics deems it appropriate to grant human drivers the freedom to deliberate and decide among these options within their unique contexts.

Indeed, it may not be appropriate to determine the best, commendable, or permissible action without considering the context. To illustrate this point, let's consider a scenario where I, a qualified driver, am accompanied by my three small children in an automated small car during a trip. Suddenly, we encounter a brake failure situation, and there are five individuals in front of our car. I am aware that the car has been programmed to swerve onto a barrier to avoid hitting them. However, I notice that next to the road barrier is a deep ditch, and if I follow the programmed algorithm, our car will inevitably veer into the ditch, endangering the lives of my children and myself.

In such a situation, I make the decision to take control of the steering wheel and keep the car on its current course to protect the lives of my children. As a result, three individuals among the five in front of us are tragically killed, while two individuals at the fringe of the crowd remain unharmed. In contrast, if I had not taken over the steering wheel and followed the programmed algorithms, all four of us would have perished. In this specific case, it becomes evident that swerving towards the barrier was not the best or commendable action to take.

Confucian teachings emphasize that human agents should seize important opportunities to exercise virtue and utilize their practical wisdom according to their own judgment, particularly when it comes to protecting their immediate family members. Since Confucian virtue ethics cannot be fully encapsulated in perfect digital algorithms to handle every possible situation, it is most reasonable to retain a steering wheel for human drivers to exert control and make decisions as they see fit.

There might be a concern that implementing suggestions C and D in East Asian regions may lead to a situation where only qualified human drivers would be interested in purchasing, renting, or using small automated vehicles. This is because they would have the option to protect themselves by activating the steering wheel, while individuals without a driver's license would not be allowed to do so for their own safety during a crisis. Consequently, these individuals, including children and many elders, may not be interested in using such vehicles at all. They might even hesitate to take a taxi that operates based on suggestions C and D due to the perceived risk to their lives in case of a brake failure.

In response, it is important to acknowledge that individuals have the right to refuse using such vehicles if they choose to do so. Society should support the availability of human-driven taxis as an alternative option for those individuals. At the same time, it is reasonable to expect that most individuals without a driver's license would still be able to use AVs in many situations, particularly on closed or semi-closed routes specifically designed for AV use. The occurrence of a brake failure in such controlled environments would be extremely rare.

Moreover, it is ethically appropriate for individuals to be motivated to continue learning how to drive and become qualified drivers, even when AVs are predominantly employed. This demonstrates their dedication to providing attentive care to their family members and exemplifies their special love for them, aligning with the fundamental concerns of Confucian virtue ethics.[18] Finally, although qualified human drivers have the ability to drive vehicles themselves, it is often convenient and comfortable for them to utilize AVs for most of their travel needs. The presence of a steering wheel in AVs allows them to take control when desired, providing a flexible and adaptable transportation option.[19]

[18] I am pleased to discover that a majority of the respondents in our survey conducted in Hong Kong and Shenzhen support the idea presented in my suggestion D (refer to Chap. 9 of this book). It is worth noting that this positive response from the survey participants aligns with my initial expectations, even though the survey had not been conducted at the time when I drafted this particular chapter.

[19] The inclusion of a steering wheel in automated vehicles, allowing individuals with a driver's license to take control when desired, does indeed raise complex legal issues regarding

Concluding Remarks

This chapter is based on Confucian virtue ethics as a resource to propose suggestions for programming AVs for handling brake failure scenarios in Confucian-influenced East Asian regions. It is important to note that these suggestions may be heuristic in societies with different mainstream moral cultures apart from Confucianism. However, it would be incorrect to assume that these suggestions are universally acceptable or equally useful for such societies.

Critics may argue that my culture-relevant approach to AV programming is invalid, asserting that ethical guidelines for AVs should be universally justifiable and applicable, independent of any particular culture. They emphasize that the truth of AV ethics should not depend on cultural factors. However, while ethical truths are not determined by any specific culture, it is flawed to assume that ethical justifications can be made without considering cultural perspectives. Ethical reasoning inherently incorporates cultural values and considerations, including those relevant to AV programming, as arguments stem from specific cultural contexts. The challenge arises when individuals embedded in different cultural contexts do not embrace the same premises and deliberations. Therefore, asserting that one's ethical recommendations for AV programming should be universally adopted across diverse cultural regions is ethically unreasonable and impractical, displaying arrogance.

It is worth noting that Confucian ethics is not exclusively for East Asians, just as Kantian deontology is not solely for Europeans. However, there is a distinction between consideration and acceptance. When Confucian scholars develop a framework based on Confucian culture to provide relevant recommendations specifically tailored to societies with a similar Confucian cultural background, it is relatively less presumptuous than recommending it to societies with different cultural backgrounds. Expecting all regions with diverse cultures to adopt identical AV algorithms to address brake failure situations is unreasonable. It is both theoretically invalid to justify the "truth" of one's accepted moral culture or

responsibility and liability. However, I have confidence that these matters can be appropriately addressed through the development and implementation of relevant laws and regulations. It would be prudent to entrust the resolution of these legal issues to colleagues in the field of law, who possess the expertise to navigate and tackle these intricate matters. Their involvement will be crucial in ensuring a comprehensive and fair legal framework for the operation and use of such vehicles.

ethical theory by concealing particular moral assumptions or convictions and practically ineffective, because people living in various cultures will not accept the superiority of one specific culture to dictate the programming of their AVs in the same manner.

As long as basic individual rights and freedoms are respected, it is legitimate and beneficial for each region to develop their AVs in accordance with their mainstream moral culture. Regions can learn from and cooperate with each other through genuine and peaceful dialogues, fostering mutual understanding.

Acknowledgements The earlier versions of this chapter were presented at an online conference in October 2022 and a workshop held at the City University of Hong Kong in November 23–25, 2023. These events were co-chaired by Chunyan Ding and myself, funded by the CityU CPAL research team. I would like to express my gratitude to the participants of these conferences for their discussions of my presentations. I am deeply thankful to the following colleagues and friends for their valuable comments, which have greatly assisted me in revising and finalizing this chapter: Jeeloo Liu, Li Lun, Xie Min, and Maximilian Johan Nilsson-Ladner.

References

Anderson, S. (2008). Asimov's 'three laws of robotics' and machines metaethics. *AI & SOCIETY, 22*(4), 477–493.

Awad, E., Dsouza, S., Kim, R., Schulz, J., Henrich, J., Shariff, A., Bonnefon, J. F., & Rahwan, I. (2018). The moral machine experiment. *Nature, 563*, 59–64.

Bennefon, J., Shariff, A., & Rahwan, I. (2016). The social dilemma of autonomous vehicles. *Science, 352*, 1573–1576.

Billioud, S., & Thoraval, J. (2015). *The sage and the people: The Confucian revival in China*. Oxford University Press.

Dubljević, V. (2020). Toward implementing the ADC model of moral judgment in autonomous vehicles. *Science and Engineering Ethics, 26*, 2461–2472.

Dubljević, V., Sattler, S., & Racine, E. (2018). Deciphering moral intuition: How agents, deeds, and consequences influence moral judgment. *PLoS ONE, 13*(10), e0204631. https://doi.org/10.1371/journal.pone.0204631

Dworkin, R. (1977). *Taking rights seriously*. Harvard University Press.

Fan, R. (2010). *Reconstructionist Confucianism: Rethinking morality after the west*. Springer.

Fan, R. (Ed.). (2011). *The Renaissance of Confucianism in contemporary China*. Springer.

Fan, R. (2021). A Confucian conception of public reason and bioethics. In L. Hon-lam & M. Campbell (Eds.), *Public reason and bioethics: The three perspectives* (pp. 93–134). Palgrave Macmillan.

Fan, R. (2024). Principlism as global bioethics: A critical appraisal from a Confucian perspective. *Dao: A Journal of Comparative Philosophy, 23*(3): 353–376.

Graham, A. C. (1991). *Disputers of the Dao: Philosophical argument in ancient China*. Open Court.

Harris, J. (2020). The immoral machine. *Cambridge Quarterly of Healthcare Ethics, 29*, 71–79.

Hursthouse, R. (2022). Virtue ethics. In *Stanford Encyclopedia of Philosophy*. https://plato.stanford.edu/entries/ethics-virtue/

Lau, D. C. (trans.) (1970). Mencius. Penguin Books.

Lau, D. C. (1979). *Confucius: The analects*. Penguin Books.

Leben, D. (2017). A Rawlsian algorithm for autonomous vehicles. *Ethics and Information Technology, 19*, 107–115.

Legge, J. (trans.) (1971). Confucian analects, the great learning and the doctrine of the mean. Dover Publications, Inc.

Li, J. (2012). *Cultural foundations of learning: East and West*. Cambridge University Press.

Liu, J. (2021). *Confucian robotic ethics*. An unpublished paper generously offered by the author for my reference as well as for the use of my teaching at the City University of Hong Kong in 2023.

Luetge, C. (2017). The German ethics code for automated and connected driving. *Philosophy and Technology, 30*, 547–558.

MIT Lab. (2022). *The moral machine*. Retrieved December 20, 2023, from https://www.moralmachine.net/

Moor, J. H. (1979). Are there decisions computers should never make? *Nature and System, 1*(4), 217–229.

Moor, J. H. (2006). The nature, importance, and difficulty of machine ethics. *IEEE Intelligent Systems, 21*(4), 18–21.

Nisbett, R. (2003). *The geography of thought: How Asians and Westerners think differently … and why*. Free Press.

Othman, K. (2021). Public acceptance and perception of autonomous vehicles: A comprehensive review. *AI and Ethics, 1*, 355–387. https://link.springer.com/content/pdf/10.1007/s43681-021-00041-8.pdf

Powers, T. M. (2006). Prospects for a Kantian machine. *IEEE Intelligent Systems, 21*(4), 46–51.

Rhim, J., Lee, J., Chen, M., & Lim, A. (2021). A deeper look at autonomous vehicle ethics: An integrative ethical decision-making framework to explain moral pluralism. *Frontiers in Robotics and AI, 8*, 1–18.

Sen, A. (1985). Well-being, agency and freedom: The Dewey lectures 1984. *Journal of Philosophy, 82*(4), 169–221.

Sen, A. (2013). *Confucianism as a world religion: Contested histories and contemporary realities*. Princeton University Press.

Tokens, R. (2012). Out of character: On the creation of virtuous machines. *Ethics in Technology, 14*, 137–149.

Van Norden, B. W. (2003). Virtue ethics and Confucianism. In B. Mou (Ed.), *Comparative approaches to Chinese philosophy* (pp. 99–121). Routledge.

Wallach, W., & Allen, C. (2009). *Moral machines: Teaching robots right from wrong*. Oxford University Press.

Watts, G. (2007). John Harris: Leading libertarian bioethicist. *The Lancet, 370*, 1411.

WHO. (2023). Road traffic injuries. Retrieved December 20, 2023, from https://www.who.int/news-room/fact-sheets/detail/road-traffic-injuries

Yang, B. 楊伯峻 (1980). *Interpreting the analects* (論語譯注). Zhonghua Shuju.

CHAPTER 5

Automated Vehicles, Privacy Protection, and Ethics: A Comparing Study on the Western Theories and Confucianism

Jianhui Li and Katharina Yu

Introduction

Automated vehicles (AVs) are a swiftly advancing technology poised to revolutionize both the transportation industry and our individual travel patterns. Offering the potential for increased safety, reduced traffic congestion, and enhanced mobility for those with disabilities or limited access

J. Li (✉)
Center for Philosophy, Politics and Economics, Beijing Normal-Hong Kong Baptist University, Zhuhai, China

School of Philosophy, Beijing Normal University, Beijing, China
e-mail: ljh@bnu.edu.cn

K. Yu
Faculty of Humanities and Social Sciences, Beijing Normal-Hong Kong Baptist University, Zhuhai, China
e-mail: Katharinayu@uic.edu.cn

C. Ding, R. Fan (eds.), *Chinese Ethical and Legal Perspectives on Automated Vehicles*,
https://doi.org/10.1007/978-981-96-1277-2_5

to transportation, AVs present substantial advantages. However, as with any nascent technology, the integration of AVs raises ethical quandaries, with the safeguarding of privacy emerging as a pivotal issue.

Primarily, AVs amass a wealth of data, including owner and passenger personal information, location, biometric data, and vehicle usage patterns (Glancy, 2012, p. 1172). These data serve a multitude of purposes, ranging from improving vehicle performance and safety to tailoring marketing strategies and enabling innovative business models. Nevertheless, these applications also carry inherent risks to privacy and autonomy. For example, AV companies collect user data like names, addresses, and payment information, posing a potential threat of data breaches. Additionally, passengers' electronic devices may store sensitive data that, if not properly secured, could be vulnerable to hacking or unauthorized intrusion.

AVs' collection of passenger travel patterns can disclose personal details like home and work locations, preferred destinations, travel habits, driving times, speeds, and stopping points (Lee & Hess, 2022). These data could be exploited for tracking and monitoring individuals, potentially infringing upon their privacy and endangering their safety. Such developments could have significant implications for civil liberties and individual autonomy.

Furthermore, AVs have sensors and cameras that capture passengers' facial expressions, voice, behavior, and conversations. These highly sensitive and personal data, without adequate privacy safeguards, pose a significant risk of misuse for malicious purposes such as identity theft, stalking, or blackmail. As AV technologies continue to advance and integrate more deeply into our daily lives, establishing robust ethical guidelines and regulations becomes paramount to safeguard privacy and uphold ethical standards (Glancy, 2012, p. 1196).

Secondly, AVs store data in onboard computers and cloud-based storage systems, which can be accessed by third parties such as manufacturers, service providers, and law enforcement agencies. Nevertheless, unauthorized access or misuse of this data can present a substantial risk to individual privacy and security.

Thirdly, although the data gathered by AVs can improve vehicle performance and safety, support personalized marketing, and drive innovation in business models, they also raise important considerations. Concerns such as targeted advertising, profiling, and surveillance have the potential to encroach upon individual autonomy and dignity. These practices emphasize the necessity for strong safeguards and regulations to uphold individual rights and freedoms.

To fortify the privacy of users in the context of AVs, it is imperative to delineate the entities that may infringe upon privacy and the methods they might employ. This identification process is pivotal in ensuring accountability, fortifying passengers' privacy, and fostering an ethical environment in the development and utilization of AVs.

There exists a multitude of potential actors who could compromise the privacy of AV users. These encompass manufacturers and service providers of AVs, governmental agencies, and malicious entities such as hackers and cybercriminals (Glancy, 2012, p. 1199). Acknowledging these potential threats is a fundamental step towards formulating effective strategies to safeguard user privacy in the rapidly evolving domain of automated vehicles.

Primarily, manufacturers and service providers of AVs collect and store data from the sensors and cameras integrated into their vehicles. Without stringent privacy protection measures in place, there is a risk of unauthorized access to this data by individuals or entities.

Additionally, governmental agencies may have the capability to access data accumulated by AVs for surveillance and monitoring objectives. The absence of clear regulations regarding the use and sharing of this data could potentially violate the privacy of individuals.

Furthermore, AVs are vulnerable to cyberattacks. If unauthorized individuals were to obtain access to the data gathered by AVs, they could potentially exploit it for nefarious purposes such as identity theft, extortion, or blackmail.

Recognizing potential violators and comprehending their tactics is a vital first stride in formulating ethical standards for safeguarding privacy. Generally, it is the duty of AV manufacturers and operators to guarantee that the data gathered by AVs is stored and employed in accordance with pertinent privacy statutes and rules. Additionally, third-party entities or individuals with access to these data must be held responsible for any unauthorized use or sharing of the information. Lastly, hackers and other malicious actors should be held accountable for any illicit or unethical misuse of the data obtained from AVs (Glancy, 2012, p. 1187).

Protecting privacy is of paramount importance globally, yet the principles governing its preservation may differ depending on cultural norms. This paper seeks to delve into privacy concerns concerning Automated Vehicles (AVs), conducting a comparative examination between Western and Confucian ethical frameworks to underscore the similarities and disparities in privacy protection within these distinct cultural settings.

Section II provides an overview of Western ethical guidelines for privacy protection and their application to AVs. Section III explores Confucian perspectives on privacy protection, primarily drawing from the principles outlined in the Book of Rites and discusses their implications for AVs. Section IV presents a comparative analysis of the characteristics of Western and Confucian privacy theories. Lastly, Section V synthesizes the key tenets of Western and Confucian privacy theories and their relevance to AVs.

Western Ethics for Protecting Privacy

To begin, we will explore the Western concept of privacy protection, a notion deeply embedded in the historical fabric of Western culture. As far back as the times of Ancient Greece, a distinction between public and private affairs was recognized. Further historical examination leads us to the foundation of medical ethics, the Hippocratic Oath, which intrinsically upholds the principle of preserving individual privacy:

> All that may come to my knowledge in the exercise of my profession or in daily commerce with men, which ought not to be spread abroad, I will keep secret and will never reveal it.

The origins of regulations protecting individual homes can be traced back to ancient Roman Law. Despite the emphasis placed by numerous philosophers and scholars on the importance of privacy, a definitive definition remains elusive. Samuel Warren and Louis Brandeis were among the early scholars who grappled with defining privacy. Their groundbreaking 1890 article, "The Right to Privacy," is now widely regarded as a cornerstone in privacy studies. They defined privacy as "being let alone" or the freedom from unwanted intrusion. Their arguments highlighted that privacy rights are inherent to human existence and deserve legal protection to uphold personal autonomy and individual freedom. They examined the potential threats posed by emerging technologies, such as photography and journalism, to these rights and advocated for a legal framework that includes the recognition of a tort for invasion of privacy (Warren & Brandeis, 1890). Building upon the foundational work of Warren and Brandeis, numerous Western scholars have sought to further refine and broaden the understanding of privacy. Notably, Alan Westin contributed significantly to this discourse. He posited privacy as a right that empowers individuals to decide if and how their information should be shared with others. He articulated this as follows:

> Privacy is the claim of individuals, groups, or institutions to determine for themselves when, how, and to what extent information about them is communicated to others. Viewed in terms of the relation of the individual to social participation, privacy is the voluntary and temporary withdrawal of a person from the general society through physical or psychological means, either in a state of solitude or small group intimacy or, when among large groups, in a condition of anonymity or reserve. (Westin, 1967, p. 7)

Westin's conceptual framework outlines four distinct dimensions of privacy: Solitude, Intimacy, Anonymity, and Reserve. Solitude pertains to the freedom from intrusive scrutiny by others. Intimacy involves the private enjoyment of a small group, cultivating an atmosphere of trust, tranquility, and open conversation among its members. Anonymity denotes the freedom from identification and observation in public settings and engagements. Reserve, conversely, stems from the inclination to restrict the disclosure of information to external parties, implicitly requiring others to recognize and honor these boundaries.

Westin delineates four distinct functions of privacy: Personal Autonomy, Emotional Release, Self-Evaluation, and Limited and Protected Communication. Personal Autonomy pertains to the deliberate wish to evade undue influence, control, or unwarranted exposure from external sources. Emotional Release signifies liberation from societal expectations, including emotional demands, minor deviations, and the management of losses and physical necessities. When encountered in solitude or within a supportive environment, this facet of privacy creates a "safe haven," liberating individuals from societal pressures and providing avenues for emotional catharsis.

Self-Evaluation encompasses the process of integrating experiences into coherent patterns and imprinting one's individuality on events. This involves tasks such as information processing, strategic planning (such as timing disclosures thoughtfully), synthesizing experiences, and creating opportunities for introspection of a moral and spiritual nature.

The last function, Limited and Protected Communication, divides into two parts: Limited Communication sets boundaries in interpersonal interactions, while Protected Communication enables the exchange of personal information with trusted individuals (Westin, 1967).

Westin contends that his theoretical view of privacy is particularly relevant to Western democracies due to its congruence with the sociopolitical principles of these societies.

Herman T. Tavani comprehensively analyses various theorists' viewpoints on privacy, categorizing them into four main types: Nonintrusion, Seclusion, Limitation, and Control theories (Tavani, 2007, p. 4). Warren and Brandeis' definition of privacy as "the right to be let alone" is interpreted by Tavani as a Nonintrusion theory. The Seclusion theory, on the other hand, equates privacy with the condition of "being alone." Tavani aligns Ruth Gavison's conceptualization with this theory, as it depicts individuals as attaining "perfect privacy" when they become "completely inaccessible to others."

The Limitation theory posits that privacy is preserved when access to personal information is restricted or confined within certain contexts. Tavani highlights Gavison's perspective on this concept, defining privacy as the "restriction of others' access to information about individuals" (Gavison, 1980, p. 428).

The Control theory of privacy asserts that privacy is manifested in a person's control over their personal information. Tavani points out that a significant group of scholars support this privacy theory. For instance, Charles Fried argues that privacy "is not simply an absence of information about us in the minds of others, rather it is the control over information we have about ourselves" (Fried, 1990, p. 54). Arthur Miller defines privacy as "the individual's ability to control the circulation of information relating to him" (Miller, 1971, p. 25). Similarly, Alan Westin describes privacy as "the claim of individuals … to determine for themselves when, how, and to what extent information about them is communicated to others" (Westin, 1967, p. 7).

Moor and Herman T. Tavani present a unique perspective on privacy, which they term it as "Restricted Access/Limited Control (RALC) theory". According to the principles of RALC, an individual possesses privacy "in a situation with regard to others [if] in that situation the individual … is protected from intrusion, interference, and information access by others" (Moor, 1997, p. 30). This theory suggests that the privacy of information is not solely predicated on an individual's absolute control over its access. Instead, it is contingent on the creation of normative boundaries intended to prevent unauthorized access.

In interpersonal interactions, there exists a certain level of control, albeit restricted, over the relationships formed. Privacy laws that safeguard data within a particular context by normatively restricting access offer individuals a measure of control. In the realm of privacy management, absolute control over information is not imperative. Instead, individuals require a certain level of control in three crucial areas: choice, consent, and rectification.

An individual must have the capacity to select situations that grant others the preferred level of access, spanning from total privacy to complete public exposure. Privacy management can also be influenced by a consent process, where one may voluntarily waive their right to restrict others from accessing specific types of personal data. In terms of correcting personal information to manage privacy, it is crucial for individuals to have access to their own data and the authority to modify it if needed. These limited controls—choice, consent, and correction—are frequently facilitated by thorough privacy policies.

Several scholars have delved into the theoretical exploration of the concept of privacy, each offering distinct viewpoints. For the scope of this discussion, we have concentrated on a specific group whose key ideas embody the fundamental principles of privacy. Primarily, these scholars examine privacy from a rights-based standpoint, viewing it as a fundamental right manifested in various forms—encompassing the rights to solitude, non-intrusion, autonomy, intimacy, anonymity, and reserve, among others.

For example, the right to solitude emphasizes an individual's need for physical and psychological boundaries, highlighting the right to control access to one's body and mind. The right to autonomy underscores an individual's entitlement to direct their own life and make decisions without external interference. The right to intimacy underscores the importance of personal relationships and the need for individuals to have a space where they can authentically express their deepest thoughts and emotions.

Essentially, privacy is viewed as an individual's entitlement to manage their personal information and to be shielded from unjustified or unauthorized intrusions into their personal affairs. Philosophically, privacy can be seen as a fundamental aspect of human dignity and autonomy. This encompasses the right to withhold specific information from others, as well as the right to determine how personal information is utilized and disclosed.

Given the necessity of personal information collection by others in various circumstances, Western privacy theories emphasize the significance of establishing boundaries on the utilization of personal data. Scholars and policymakers have developed numerous principles governing the use of personal data. For instance, the principle of data minimization advocates that organizations should only collect the essential data required for a specific purpose, avoiding irrelevant or excessive data. The principle of purpose limitation asserts that personal data should be utilized solely for the purpose for which it was collected, and not for other purposes without the individual's explicit consent. The principle of data accuracy underscores the paramount importance of ensuring the accuracy, up-to-date nature, and relevance of personal data to the intended purpose.

Western privacy theories also highlight the importance of accountability in cases of privacy violations. This can be achieved through the establishment of legal and regulatory frameworks that impose penalties for unauthorized access, use, or disclosure of personal data. Additionally, ethical frameworks can promote responsible data practices and encourage organizations to prioritize the privacy and security of personal data. Consequences for breaching the privacy of AVs data may include fines, legal repercussions, and damage to the reputation of organizations and individuals who violate privacy laws and ethical standards. Western privacy theories stress the significance of individual rights, freedoms, and autonomy, set limits on the use of personal data, and advocate for accountability in cases of privacy breaches. These principles could play a crucial role in shaping the development and implementation of AVs, ensuring that individual privacy rights are upheld and protected. Here are some of the essential principles of Western privacy protection applicable to AV users:

Informed Consent: The principle of informed consent posits that individuals should have control over their personal data, and they must be adequately informed and give their consent before any collection or usage of their personal data. Regulatory policies for AVs should prioritize individual privacy rights protection, potentially including mandates for explicit user consent before data collection or sharing.

Purpose Limitation: This principle asserts that personal data should only be collected and utilized for specific, pre-defined purposes. There should be stringent limitations on the nature of data that can be collected and the ways it can be used.

Data Minimization: According to this principle, only the least amount of personal data necessary for the purpose should be collected and used.

Transparency: This principle grants individuals the right to be informed about the collection and usage of their personal data.

Responsibility and Accountability: Entities that collect and use personal data are responsible for its protection and must be held accountable for any misuse or security breaches. AV companies and other organizations handling personal data are perceived as having an obligation to protect that information and use it solely for legitimate purposes. This may include mandates for data security and the establishment of effective safeguards against unauthorized access or misuse of personal data.

Privacy Protection from Confucian Perspective

Confucianism, an ethical and philosophical system originating from China, prioritizes the significance of interpersonal connections, societal equilibrium, and ethical conduct. Numerous East Asian territories, including the Chinese Mailand, Taiwan, and Hong Kong, persist in their adherence to Confucian moral culture. Although Confucianism is no longer the official political ideology in the Chinese Mailand, it significantly influences individuals' ethical conduct. While Confucianism does not present a structured privacy theory, it encompasses various concepts aimed at safeguarding personal privacy. As the famous professor for studying Chinese theory of privacy Bonnie S. McDougall says:

> One of the most prevalent myths in the West about Chinese culture is the one that goes 'The Chinese don't have a sense of [or, a word for] privacy'. ... There is also a widespread perception within China that privacy is a concept introduced into China from the West since the nineteenth century. Scarcely less prevalent is the notion that if the Chinese do have a sense of privacy, or a word for privacy, it must be a single one. None of these propositions is true. (McDougall, 2004, p. 1)

Famous Chinese sociologist Xiaoqiang Feng also says in his article:

> Many people believe that Confucianism has no concept of privacy because Confucianism emphasizes collective values and lacks a strict concept of individual rights, leading to a lack of a modern sense of privacy. Some even think that the existence of rituals and customs exacerbates this indifference to

> privacy. However, if one carefully examines the Confucian classics, it is not difficult to find that these views are not correct. (Feng, 2020)

The Chinese term for privacy, "yinsi" (隐私), is composed of two separate characters: "yin" (隐) and "si" (私). "Yin" translates to "concealed" or "hidden," often carrying a negative connotation associated with illicit affairs. Conversely, "si" signifies "private" or "that which is not intended for public disclosure," encompassing both the notions of "private" and "privacy." "Si" (private/privacy) has long been an antonym to "gong" (public or common) or "guan" (official). Regardless of whether the term is used independently or within a compound, it typically carries a negative undertone in most contexts, suggesting dishonorable activities performed in secret and/or driven by disreputable motives. In ancient China, "gong" held significant importance. "The Great Harmony" is a chapter title in the "Li ji" (Book of Rites), attributed to Confucius. It commences with "Tian xia wei gong," which literally translates to "under heaven, everything is common." "Gong" is typically interpreted as "the common good," as in "All under Heaven was utilized for the common good." The subsequent text shows the importance of "gong" in Chinese culture:

> Men did not love their parents to the exclusion of the parents of others, nor their own sons to the exclusion of others … Not wishing to be wasteful of their possessions, nevertheless they did not keep them for purely personal use; not wishing to be inactive in the application of their strength, at the same time they did not exert it merely in their own behalf. (Dai, 2017, p. 419)

Though "gong" is seen as a very important value in ancient China, it does not mean there is completely no space of "private". As McDougall says:

> In traditional Confucian thinking, the state and the family operated in largely separate realms, each, in principle, supporting the other, while the conflicts that in practice arose between them were portrayed in historical and literary writing as moral tragedies of the highest order. Privacy in traditional China resides primarily in the family unit as distinct from the state, but literary evidence also shows the existence of private realms for individuals, couples and non-familial small groups. (McDougall, 2004, p. 3)

China has a long history of civilization. In ancient time, civil disputes were resolved by moral principles of Confucianism, called li (礼, ritual or rite). Consequently, with the establishment of rites, privacy was indirectly protected to some extent. "The core of theory of li is to cultivate morality.

In Confucians' view, both ideal social order and perfect human relations can be achieved on the basis of theory of li (morality)" (Wang, 2012, p. 137).

In the Analects, Confucius once said: "Do not watch if it contradicts with li; do not listen if it contracts with li; do not speak if it contradicts with li; and do not act if it contradicts with li (Confucius, 2015, p. 89)."

The Chinese perspective on privacy is intricately connected to li. The Book of Rites serves as an illustrative example, as it encompasses a wealth of rituals aimed at preserving human privacy. These rituals establish a bedrock for privacy protection by underscoring the significance of personal relationships and individual autonomy. Some notable excerpts include:

> (1) "When you go up to the hall of the host, your voice must be raised to show there is a guest coming. If there is a pair of shoes outside of the door, enter when you hear there are voices of talking, and do not enter when you hear no voice. When you enter a door, you must lower your gaze. When you enter a door, your hands should put at front of your chest like lifting the latch, and you should look straight ahead and do not look back. If the door is open, leave it open; if it is closed, leave it closed. If there is someone else entering afterward, close the door slightly but do not latch it." (Dai, 2017, p. 17)

The passage lays out a series of behavioural norms for entering a host's room or hall, accentuating the significance of proper decorum. This includes audibly announcing oneself, directing one's gaze downwards when crossing a threshold, and refraining from scanning the room upon entry. These guidelines reflect a deep-rooted understanding of privacy and its interplay with ritual.

In this context, privacy fundamentally represents personal space, going beyond just physical boundaries to include a sense of spiritual exclusivity. It promotes individual autonomy in interactions with others, requiring a regard for one's personal limits. The customs linked to entering a host's room or space—like audibly announcing oneself, lowering one's gaze, and refraining from scanning the area—serve as demonstrations of respect for the host's privacy.

Within the Confucianist framework, privacy is intrinsically connected to appropriate etiquette or rituals in interpersonal interactions. These interactions establish boundaries that recognize certain spaces as personal and inviolable, preventing others from encroaching upon them. Notably, both privacy and ritual share similar boundary structures, as they both pertain to interpersonal communication and the regulation of societal norms.

> (2) "When accompanying a guest, the host should always give way to the guest at each door. When the guest arrives at the bedroom door, the host should enter first, lay out the seats, and then go out to welcome the guest going in. If the guest repeatedly excuses himself, then the host should go ahead and lead the guest into the bedroom." (Dai, 2017, p. 18)

The provided excerpt elucidates the intricate etiquette that governs the reception of a guest in a host's residence. This etiquette serves multiple purposes: it not only conveys respect but also safeguards the host's privacy. Unlike the process of entering the exterior door, the host is the first to enter the bedroom. This step enables the host to verify that there are no personal items within the bedroom that may be inappropriate for the guest to observe. Similarly, the guest's repeated refusal to enter the bedroom reinforces this principle. These actions emphasize the delicate equilibrium between extending hospitality to guests and preserving the host's privacy.

The aforementioned accounts of etiquette for receiving and greeting guests subtly intertwine the concept of 'privacy'. They suggest an inherent cognizance and corresponding reverence for private space. The instances highlighted, such as the audible announcement prior to entering the hall and the guest's refusal to intrude upon the host's personal space, exemplify an understanding and respect for individual privacy. These accounts demonstrate that the notion of privacy held substantial value in traditional Chinese culture and was embodied in their societal etiquette.

Beyond personal space, the concept of privacy encompasses the preservation of personal intentions. At its essence, this involves the refraining from public disclosure of personal intentions, thereby fostering an expectation for others to arrive at an unspoken understanding. Indeed, the protection of personal intentions is intrinsically connected to the privacy of personal space, with the latter serving as the foundational rationale for the former. The recurrent emphasis on respecting personal intentions within the context of privacy is highlighted in the "Book of Rites." For instance:

> (3) "Walking behind your teacher, you should not run to the other side of the road to talk to others. When you meet a teacher on the road, you should step forward quickly and stand upright with your hands clasped. If the

teacher talks to you, you should answer; if the teacher does not talk to you, you should quickly step back." (Dai, 2017, p. 16)

This paragraph shows that you should not only behave politely to your teacher, but also respect your teacher's personal intention.

(4) "When assisting a gentleman, if someone comes in and says, 'if you have time, I have something to report,' the assistants should move to the side and wait." (Dai, 2017, p. 25)

This passage highlights that servants or assistants are expected to refrain from eavesdropping on private conversations that a gentleman may engage in with his guest or friend. This practice further emphasizes the importance of privacy and discretion within interpersonal dynamics.

(5) "If you see two people sitting together, or two people standing together, don't squeeze there anymore. If you see two people standing next to each other, don't walk between them." (Dai, 2017, p. 27)

This paragraph illustrates that it's important to respect another person's private space and not to intrude upon their personal intentions.

(6) "If you go to visit your father's good friend, and he does not let you enter the door, you dare not make your own decision to enter; if he doesn't let you go backward, you dare not make your own decision to go backward; if he doesn't ask questions, you dare not speak to him. These are all qualities that a filial son should have." (Dai, 2017, p. 12)

This paragraph elaborates that a filial son should respect his parents' friend and respect their personal space and personal intention.

(7) "The words of men do not enter the boudoir, and the words in the boudoir do not spread outside." (Dai, 2017, p. 27)

Throughout historical periods, societal roles were frequently divided along gender lines, with men occupying positions of public authority and women confined to the domestic sphere. As outlined in the Book of Rites, this societal norm suggests that men were discouraged from sharing their professional knowledge within the household, while women's private conversations were not to be shared in public. This distinction represents an early Chinese distinction between public and private spheres.

The Book of Rites, through various examples, emphasizes the paramount importance of safeguarding personal privacy in diverse aspects of societal, everyday, and familial life. These rites propose that four fundamental privacy concepts were recognized in ancient China: personal space, personal intention, personal information, and personal domains. In the Confucian perspective, the protection of privacy is not perceived as an individual right, but rather as a virtue that conveys respect for others. Consequently, the Confucian interpretation of privacy possesses the following characteristics:

Firstly, the cultivation of virtues that respect others' privacy is encouraged. The Book of Rites posits that respecting others' privacy is not only a moral virtue but also a crucial element in fostering a harmonious society.

Secondly, the development of virtues that respect privacy is achieved through adherence to ritual practices and norms of etiquette. The Book of Rites is replete with rules and norms pertaining to etiquette, including those that relate to privacy protection. These norms guide behavior in various situations, such as entering another person's home or navigating social interactions.

Thirdly, the Book of Rites emphasizes the significance of sincerity and honesty. It asserts that an individual's words and actions should align with their genuine thoughts and feelings, and they should avoid deception or obscuring the truth. Furthermore, Confucian philosophy promotes the concept of "Shendu" (慎独, self-discipline in solitude), which refers to the heightened self-awareness individuals exhibit when unsupervised, adhering to moral norms, and refraining from actions that contravene their moral convictions or humanistic principles. Exercising caution in solitude represents a form of self-discipline, conveys a particular sentiment, signifies a level of personal cultivation, and embodies a form of openness and sincerity.

Fourthly, within the societal hierarchy, deference is accorded to individuals with elevated social status, such as teachers, parents, and elders. This respect extends to the sanctity of their private spaces and lives. The Book of Rites strongly emphasizes the significance of this respect as a foundational element for maintaining harmony and stability within familial and societal structures.

Fifthly, the cultivation of trust is fundamental to the rituals associated with privacy protection. Trust, a cornerstone value in Confucianism, is frequently founded upon adherence to Confucian rituals. When individuals or entities effectively uphold these customary practices, a foundation of trust is established among community members.

Sixthly, the enhancement of social well-being and harmony is a potential outcome of privacy protection. This can be achieved by ensuring that personal data is utilized for beneficial purposes, such as improving vehicle performance and safety, while concurrently minimizing risks to individuals' privacy and security.

The Confucian perspective on privacy protection offers a valuable framework for the formulation of policies that strike a balance between the benefits of autonomous vehicle technology and the necessity to protect user privacy and foster social harmony. To achieve this equilibrium, companies within the autonomous vehicle industry must cultivate a culture of trust.

In the domain of AVs, it is imperative for companies to establish robust security systems that ensure the safe collection, storage, transfer, and extraction of data. The process of information collection should be conducted in an anonymized manner, while the storage and transfer of this data should be encrypted. The information system should be fortified to a degree that resists unauthorized breaches. Furthermore, the AV company should be held accountable for any potential data leaks that compromise the privacy of its customers.

Moreover, it is essential to uphold the privacy of AV users when gathering information. Personal data, such as the name of the owner or passenger, ID card information, audio and video recordings from within the AV, and details of the vehicle's travel and stops, should only be collected with explicit consent from the owner or passenger. In particular, any internal recording systems in the AVs should be designed to give users the option to turn them on or off. This guarantees that the collection of personal information is a decision made by the AV users.

In addition, the utilization of gathered information should be contingent upon the informed consent of AV users. This principle pertains not only to the collection of data but also to the subsequent use of that data. If a user does not provide consent, the use of their information must be stopped.

Guided by these principles, AV manufacturers should devise specific guidelines or measures to protect the privacy of AV users. These guidelines may encompass:

Transparency: AV manufacturers should furnish users with clear and concise information about the type of data being collected and its intended use. This ensures that individuals are cognizant of the implications of using AVs and can make informed decisions about their privacy.

Collective Informed Consent: In keeping with Confucian tradition, it is suggested that informed consent should be obtained from the family or collective group (Fan, 2015). This reflects the Confucian emphasis on the primacy of family and collective groups over the individual. Hence, it is preferable to involve the family or collective group in deciding whether to grant permission for data collection and use.

Anonymization: AV manufacturers should anonymize personal data by either removing identifiable information or aggregating data. This strategy can safeguard individuals' privacy and reduce the risk of data breaches or misuse.

Encryption: AV manufacturers should employ encryption techniques to shield personal data from unauthorized access or use. This ensures that personal data is accessible only to authorized parties, thereby minimizing the risk of data breaches or misuse.

Accountability: AV manufacturers should ensure that responsible parties are held accountable for the collection, storage, and use of personal data. This can guarantee that individuals' privacy is respected and protected, and that responsible parties are held liable for any breaches or misuse of personal data.

Comparing the Western Theory with the Confucian View

A comparative analysis of Western and Confucian theories of privacy reveals both convergences and divergences.

In terms of similarities, both cultural perspectives prioritize human privacy and its protection. For instance, the concept of personal space is treated with respect in both cultures. Western theories advocate for the right to solitude, while Confucianism encourages respect for another person's private space. Both cultures also underscore the importance of acknowledging personal intentions. Western theories emphasize individual autonomy, while Confucianism highlights respect for an individual's will. Furthermore, both perspectives advocate for the protection of personal information. In Western theories, personal information control lies with the individual to whom the information pertains. In contrast, Confucianism discourages unauthorized access to another person's personal information.

Despite their shared goal of protecting individual privacy, the theoretical underpinnings of these perspectives diverge substantially. Western theories of privacy are predominantly rights-based, while Confucianism adopts a virtue-based approach. Rights-based theories assert that privacy is an inherent human right, whereas Confucianism posits that respecting privacy is a virtue that individuals should cultivate towards others. It is worth noting that while there are interest-based and virtue-based Western theories of privacy, they do not constitute the prevailing narrative. Consequently, this analysis primarily focuses on comparing rights-based Western theories with Confucianism.

The methodologies for privacy protection also differ between the two cultures. Western theory identifies numerous scenarios for privacy protection, including solitude (freedom from observation), intimacy (seclusion within a small group for relaxed interaction), anonymity (freedom from identification and surveillance in public), and reserve (limiting personal disclosures). In contrast, Confucianism advocates for privacy protection through the practice of ritual etiquette. By adhering to established norms of etiquette and moral values, privacy is inherently protected.

Furthermore, the functions of Western and Confucian theories diverge. Western theory seeks to safeguard personal autonomy, facilitate emotional release, enable self-evaluation, and foster limited and protected communication. Conversely, Confucianism aims to promote respect for others and establish a harmonious society. To achieve these objectives, Western theory posits privacy as a universal human right, while Confucianism views privacy awareness as a ritual virtue that a cultivated individual should possess. In Confucianism, individuals are expected to prioritize communal and societal needs over personal rights and interests. Consequently, personal privacy protection is fundamental to maintaining social harmony and stability.

Finally, the application of privacy norms varies between Western and Confucian cultures. In Western theory, privacy rights are universally applicable, irrespective of social status. In contrast, Confucianism prioritizes the privacy of individuals with higher social status, particularly teachers, parents, and elders. Confucianism views the family as the societal foundation and values the relationships and obligations within it. Consequently, the Confucian model of informed consent differs from the Western model. In the Confucian model, consent is a family decision, with the ideas of elder members holding considerable weight.

Conclusion

Addressing privacy protection in AVs requires a comprehensive strategy. Western and Confucian perspectives on privacy, while both emphasizing the importance of privacy protection, diverge in their approaches and goals. Western theories consider privacy a basic human right deserving of safeguarding. In contrast, Confucianism regards privacy protection as a virtue embedded in societal customs. In a Confucian community, observance of societal customs is customary, with privacy protection being one of these customs. While Western privacy theory aims to uphold human rights, Confucian philosophy aims to promote social harmony.

These differing perspectives on privacy protection can inform the development of privacy protocols in the realm of AVs. Both theories agree that AV users should be informed about and consent to the gathering and utilization of their data. However, Confucianism, highlighting the importance of social harmony, recognizes that consent in the context of AV usage may at times necessitate familial agreement, particularly when the users are part of a family unit. In this context, the decisions of parents hold considerable influence due to the elevated reverence they receive in a Confucian community.

This comparative examination underscores the importance of cooperation among policymakers, regulators, and industry participants in creating privacy protection standards and frameworks that align with these principles. These standards should also be flexible enough to accommodate advancements in technology and shifts in societal norms. Embracing a holistic approach to privacy protection will guarantee the safety and functionality of AVs for all users.

References

Confucius. (2015). *Analects*. Beijing United Publishing Ltd.

Dai, S. (Ed.). (2017). *Book of rites*. Zhonghua Book Company.

Fan, R. (Ed.). (2015). *Family-oriented informed consent*. Springer.

Feng, X. (2020). The concept of privacy in Confucian tradition. *Tianfu Luntan*, No. 6.

Fried, C. (1990). Privacy: A rational context. In M. David Ermann, M. B. Williams, & C. Guitierrez (Eds.), *Computers, ethics, and society* (pp. 50–63). Oxford University Press.

Gavison, R. (1980). Privacy and the limits of the law. *Yale Law Journal*, *89*, 421–471.

Glancy, D. J. (2012). Privacy in autonomous vehicles. *Santa Clara Law Review*, *52*(4), 1171–1239.

Lee, D., & Hess, D. J. (2022). Public concerns and connected and automated vehicles: Safety, privacy, and data security. *Humanities & Social Sciences Communications*, *9*, 90.

McDougall, B. S. (2004). Privacy in Modern China. *History Compass* 2 AS 097, 1–8.

Miller, A. (1971). *The assault on privacy*. Harvard University Press.

Moor, J. H. (1997). Towards a theory of privacy in the information age. *Computers and Society*, *27*(3), 27–32.

Tavani, H. T. (2007). Philosophical theories of privacy: Implications for an adequate online privacy policy. *Metaphilosophy*, *38*(1), 1–22.

Wang, H. (2012). The conceptual basis of privacy standards in China and its implications for China's privacy law. *Frontiers of Law in China*, *7*(1), 134–160. https://doi.org/10.3868/s050-001-012-0007-4

Warren, S., & Brandeis, L. (1890). The right to privacy. *Harvard Law Review*, *4*(5), 193–220.

Westin, A. F. (1967). *Privacy and freedom*. Atheneum Press.

PART III

Chinese Legal Accounts

CHAPTER 6

Navigating the Evolving Risk Landscape of Automated Driving: Regulatory Approaches and Challenges

Ying Wang

As society hurtles towards an era of unprecedented technological advancement, the emergence of automated driving systems stands as a pivotal milestone. This technological leap promises a paradigm shift in transportation, offering enhanced safety, efficiency, and convenience, but also inducing unpredicted harms and risks. Over the past decades, regulatory bodies and legislators worldwide have endeavored to grapple with the complexities of integrating autonomous vehicles into the fabric of modern transportation systems. As automated driving systems continue to advance in sophistication and capability, existing legislation and regulations designed for conventional vehicles have struggled to keep pace with the evolution of autonomous technologies.

Automated driving technologies continue to advance at a rapid pace, driven by ongoing research and development efforts in both industry and

Y. Wang (✉)
Law School, Renmin University, Beijing, China
e-mail: yingwangrenmin@ruc.edu.cn

C. Ding, R. Fan (eds.), *Chinese Ethical and Legal Perspectives on Automated Vehicles*,
https://doi.org/10.1007/978-981-96-1277-2_6

academia. Recent breakthroughs in machine learning, sensor technology, and artificial intelligence have propelled automated driving closer to widespread deployment. In the U.S., self-driving car companies Waymo and Cruise has been allowed to deploy driverless Robot taxi in San Francisco since August, 2023.[1]In China, leading automated driving company Baidu Apollo has achieved commercial operation and testing of automated driving without human intervention inside the vehicle in multiple Chinese cities including Beijing, Chongqing, and Wuhan. With support and guidance from national and local governments, Baidu's Apollo L4 automated driving system testing has accumulated over 90 million kilometers. As of January 2, 2024, the total service volume of its Robo-Taxi platform has exceeded 5 million rides.[2] The accelerated commercialization of autonomous driving applications also brings about more regulatory and legal challenges.

Introduction: Evolving Automated Driving Technology and Changing Risk Landscape of Intelligent Transportation

Automated driving, empowered by the emerging technologies of the Fourth Industrial Revolution,[3]comprises a diverse set of emerging concepts that not only involve innovative individual mobility but also general trend toward automation and connectivity in the era of big data and AI.

The Evolution of Automated Driving Technology and Classification of Automation Levels by Industrial Standards

Early back to 2012, definitions of vehicle automation-degrees was put forward by the German Federal Highway Research Institute: Full automation-the system takes over longitudinal and lateral control completely and permanently. In case of a takeover request that is not

[1] Greg Bensinger, San Francisco moves to center of robotaxi universe after California agency vote. https://www.reuters.com/business/autos-transportation/california-agency-vote-san-francisco-robotaxi-expansion-amid-heavy-opposition-2023-08-10/

[2] Jiang Haoran, Speed up the improvement of the legal system for autonomous driving. https://baijiahao.baidu.com/s?id=1792949937272591749&wfr=spider&for=p

[3] See Klaus Schwab, The Fourth Industrial Revolution: What It Means and How to Respond, in: Foreign Affairs, December 12, 2015, https://www.foreignaffairs.com/world/fourth-industrial-revolution

followed, the system will return to the minimal risk condition by itself; High automation-The system takes over longitudinal and lateral control; the driver is no longer required to permanently monitor the system. In case of a take-over request, the driver must take-over request, the driver must take-over control with a certain time control with a certain time buffer; Partial automation-The system takes over longitudinal and lateral control, the driver shall permanently monitor system and shall be prepared to take over control at any time; Driver Assistance-The driver permanently controls either longitudinal or lateral control. The other task can be automated to a certain extent by the assistance system; Driver Only-Human driver executes manual driving task (Gasser & Westhoff, 2012).

Two years later, the International Society of Automotive Engineers' (SAE) issued its first edition of Levels of Driving Automation,[4] Deemed as capturing the emerging descriptive consensus about the comprehensive technologies most systematically,[5] it has become the most internationally impactful standard of levels of driving automation. In recent years, with the steady advancement of technologies like the Internet of Things, precise positioning systems, machine learning algorithms, and artificial intelligence vision, automated driving technology, particularly advanced automated driving technology beyond L3 level, has experienced remarkable development. The SAE J3016 Standard was updated in 2021 to align with the evolving technologies and deployment strategies.[6] While maintaining its 6 driving automation level structure—No Automation (Level 0), Driver Assistance (Level 1), Partial Automation (Level 2), Conditional Automation (Level 3), High Automation (Level 4) and Full Automation (Level 5), the 2021 SAE J3016 Standard introduced two distinct remote support function concepts: remote assistance and remote driving and clarified further the differences between SAE Level 3 and SAE Level 4.[7] Given that only at Level 3 and above of automated driving levels can automated driving systems effectively supplant human drivers in performing driving

[4] See Taxonomy and Definitions for Terms Related to On-Road Motor Vehicle Automated Driving Systems J3016_201401, https://www.sae.org/standards/content/j3016_201401/

[5] See OECD International Transportation Forum: Automated and Autonomous Driving Regulation under uncertainty, Corporate Partnership Board Report, https://cyberlaw.stanford.edu/fication/files/15CPB_AutonomousDriving.pdf

[6] See Taxonomy and Definitions for Terms Related to Driving Automation Systems for On-Road Motor Vehicles J3016_202104, https://www.sae.org/standards/content/j3016_202104/

[7] https://www.sae.org/blog/sae-j3016-update

tasks (simultaneously engaging in lateral and longitudinal driving maneuvers while undertaking environmental monitoring), the concept of automated driving is progressively being circumscribed to these levels. To ensure clear delineation of the levels, SAE and ISO are clarifying that the initial three levels (L0, L1, and L2) should be termed as "Driver Support Systems", whereas L3, L4, and L5 are reserved for authentic "automated driving systems".

The classification framework for driving automation introduced by SAE J3016 has garnered significant attention and implementation within the global automotive sector, including in China. The continuously maturing industrial model through demonstration applications and calls for mass production and commercial deployment, of products urges China to establish indigenous classification standards tailored to the unique needs and contexts of the Chinese automotive landscape.[8] In August 2021, China issued the Chinese National Taxonomy of Driving Automation for Vehicles also impacted by The SAE J3016 Standard also introduces similar six Levels of automation[9]: Emergency Assistance (应急辅助，Level 0); Systems can detect certain events and provide limited assistance but do not control vehicle movement continuously; Partial Driver Assistance (Level 1, 部分驾驶辅助); Systems assist with either lateral or longitudinal vehicle control under specific conditions, but drivers must supervise and intervene when necessary; Combined Driver Assistance (Level 2, 组合驾驶辅助); Systems control both lateral and longitudinal vehicle movements under specific conditions, with drivers required to monitor and intervene as needed; Conditionally Automated Driving (Level 3, 有条件自动驾驶); Systems can perform all driving tasks under specific conditions but require human backup for certain situations or failures; Highly Automated Driving (Level 4, 高度自动驾驶); Systems can handle all driving tasks under specific conditions and mitigate risks if the system fails to continue driving; Fully Automated Driving (Level 5, 完全自动驾驶); Systems can operate the vehicle under all conditions without human intervention.

The SAEJ3016 Standard nominates the specific term for a Level 3 to 5 system as "automated driving systems", in contrast to the generic term of "driving automation system" for any Level 1 to 5 system, while the

[8] See "The formal release of GB/T 40429–2021 Classification of Automotive Driving Automation" http://www.catarc.org.cn/xinwen/show-3334.html

[9] Chinese National Taxonomy of Driving Automation for Vehicles. https://openstd.samr.gov.cn/bzgk/gb/newGbInfo?hcno=4754CB1B7AD798F288C52D916BFECA34

Chinese National Standard categorizes Level 1 and Level 2 more explicitly as "Assisted Driving" and Levels 3 to 5 as "Automated Driving". The Drafter of the Chinese National Standard for Automated Driving believes that this delineation helps accurately convey the vehicle's automation level and enhances the public's precise understanding of automation levels.[10]

The British Standards Institute (BSI) has developed publicly accessible specifications (PASs), a type of fast-tracked standard, outline best practices, related to self-driving vehicles in collaboration with industry partners, local authorities, and academia. As noted by the UK regulator, specifications and standards offer detailed technical or operational guidance, complementing regulations by aiding compliance or serving as evidence of industry adherence to best practices. Standards are also more adaptable to changes in technology, practices, or terminology, making them particularly valuable for rapidly evolving technologies like self-driving vehicles.[11]

Technological Convergence and the Shifting Risk Landscape

In Chinese legislative and regulatory terminology, automated vehicles or self-driving cars are usually referred as the intelligent and connected vehicle (智能网联汽车，ICV), that is, a vehicle which is equipped with on-board sensors, CPU and other devices, facilitating information communication and data sharing between the vehicle and the outside for decision-making in navigation, movement and collision avoidance. It has the functions of complex environment perception, intelligent decision-making, cooperative control, etc., and is able to drive safely, efficiently, comfortably and energy-savingly, and ultimately realizes a new generation of vehicles that can replace human driving operation.[12] Although not

[10] National Technical Committee of Auto Standardization, http://www.catarc.org.cn/xinwen/show-3334.html

[11] Connected & Automated Mobility 2025: Realising the benefits of self-driving vehicles in the UK, 2022 August, https://assets.publishing.service.gov.uk/media/62ff438c8fa8f504cdec92df/cam-2025-realising-benefits-self-driving-vehicles.pdf

[12] See Energy Saving and New Energy Vehicle Technology Roadmap Strategic Advisory Committee, China Automotive Engineering Association, Energy Saving and New Energy Vehicle Technology Roadmap of China Society of Automotive Engineering, China Machinery Press, 2016, p. 189.

widely adopted in other jurisdictions, this terminology aptly encapsulates the principal technological attributes of next-generation vehicles. The most essential technologies utilized at least includes (Li et al., 2017, p. 3; Li & Wang, 2016, p. 5):[13]

- Environmental perception technology: the utilization of machine vision for image recognition, the application of radar (utilizing laser, millimeter-wave, and ultrasonic frequencies) for surrounding obstacle detection, multi-source information fusion technology, and redundant sensor design techniques.
- Vehicle-to-X connectivity (V2X): V2X technologies: using of wireless technologies to achieve real-time two-way communication among vehicles (V2V) and between vehicles and infrastructure (V2I).
- Decision and control algorithms: decision, planning and control algorithms for a cooperative, safe, human compatible traffic automation.
- Digital infrastructure: static and dynamic digital representations of the physical world with which the automated vehicle will interact to operate.

According to the German scholar Hilgendorf, the convergence of technologies means also the integration of different legal aspects concerning these technologies which highlights the following four concepts: "automation (Automatisierung which means free from human control), network (Vernetzung, which means smooth data flow in the overall network), caring (Fuersorglichkeit, which means technical systems understand the user preference and respond positively to it) and universality (Uebiquitaet, which means the corresponding technology is available almost everywhere) … As a result, road traffic is also a part of the 'Internet of Things' (Internet der Dinge)." (Hilgendorf, 2016, p. 87) More specifically, it will result in the diversification of involved stakeholders, including vehicle manufacturers, module and function providers of automated driving systems, drivers, V2X network service providers, among others. This diversification extends the chain of causality and blurs the boundaries of responsibility. Undoubtedly, the application of automated driving

[13] OECD International Transportation Forum: Automated and Autonomous Driving Regulation under uncertainty, Corporate Partnership Board Report, https://cyberlaw.stanford.edu/fication/files/15CPB_AutonomousDriving.pdf

technologies will reshape the risk landscape of road safety, transitioning the focus from traditional physical risks resulting from human error to a more intricate and multifaceted risk scenario. Operating as "computers on wheels," automated vehicles possess high-speed operation capabilities and advanced data transmission and processing abilities. Consequently, they are subject to both traditional mechanical and physical risks inherent in the industrial era and data privacy or security risks, including potential algorithmic harm in the era of big data and AI. Furthermore, these technologies could become intertwined, posing a new form of traffic safety risk, as noted by NHTSA:[14]

Automated Driving presents new and unique risks to motor vehicle safety because they fundamentally alter the nature of motor vehicles and their operation. Potential safety issues with vehicles operated with automated driving systems include the design and performance of sensors and other technology used to determine the vehicle's location and to identify, classify, and position other roadway users and objects. Likewise, potential safety issues may arise from decisions by software algorithms that analyze data inputs in real time to determine the appropriate vehicle response. Safety issues may also arise from the operational design domain for the ADS, and the continuing evolution and modification of these systems through software updates (including over-the-air updates).

As the technology of autonomous driving vehicles continues to evolve and improve, the aforementioned spectrum of risks is also dynamically developing, thereby presenting challenges to legal regulations. By selectively examining and exemplifying the regulatory approaches of China, Germany, the U.S., and the U.K., we aim to navigate the evolving risk landscape of automated driving. We will delve into some key questions that define the intersection of law, technology, and safety in autonomous vehicles: Does the advancement of autonomous driving technology imply technical autonomy, thereby granting the automated vehicle a regulatory autonomous status? Can the progression from L2 to L3 and L4 SAE levels gradually resolve the human-machine interface problem?

[14] NHTSA, Second Amended Standing General Order Incident Reporting for Automated Driving Systems (ADS) and Level 2 Advanced Driver Assistance Systems (ADAS), https://www.nhtsa.gov/sites/nhtsa.gov/files/2023-04/Second-Amended-SGO-2021-01_2023-04-05_2.pdf

Regulation Route Map: Navigating from Road Testing to Commercial Deployment

Generally, legal regulations governing automated driving typically follow a trajectory from road testing to commercial deployment. It is widely known that the calibration of autonomous driving algorithm models requires a significant amount of data. Alongside the simulation scenarios provided by virtual driving platforms, real-world driving data is essential for ongoing technical enhancements. Consequently, road testing emerges as a pivotal aspect in the advancement and eventual commercial deployment of automated driving technology. The refinement of automated driving capabilities relies heavily on an abundance of driving scenario inputs for continuous evolution and iterative upgrades. To ensure the safe and reliable operation of vehicles within complex traffic environments, thorough testing and verification protocols, encompassing simulation tests, controlled field tests, and real-world road tests, must be diligently conducted.[15]Road testing involves adapting vehicle performance, reliability, and durability to real-world traffic conditions prior to widespread deployment—a process essential for perfecting the automated driving systems and calibrating it to actual road conditions.

The United Kingdom has created a comprehensive regulatory regime controlling the AV Testing on public roads. The UK recently updated its code of practice: guiding automated driving trialling to support and promote the safe trialling and use of automated vehicle technologies and build public confidence in automated vehicle technologies and services.[16] The UK Government is also committed to support the industry's move towards commercialization through a dual approach, which involves reducing the risks associated with early commercial deployments and services, as well as enhancing the capabilities of the UK's supply chain. A key aspect of this strategy will be the government's ongoing commitment to ensuring that its innovation and regulatory programs are seamlessly

[15] See Management Code for Road Testing and Demonstration Application of Intelligent Connected Vehicles (Trial) (Draft for Comments), http://www.gov.cn/zhengce/zhengceku/2021-08/03/content_5629199.html

[16] See Code of Practice: Automated Driving Trialling, https://www.gov.uk/government/publications/trialling-automated-vehicle-technologies-in-public/code-of-practice-automated-vehicle-trialling

integrated and aligned in terms of both timing and objectives.[17] In November 2023, the UK government published the Automated Vehicles (AV) Bill, which aims to establish a comprehensive liability framework to ensure the safe deployment of self-driving vehicles. If passed, it will provide crucial consumer protection and pave the way for the commercialization of self-driving technologies in the UK.[18]

In the U.S., numerous technology companies and traditional vehicle manufacturers are actively testing fully automated vehicles, both in controlled environments like test tracks and virtual simulations, as well as on public roads. The National Highway Traffic Safety Administration (NHTSA) launched the Automated Vehicle Transparency and Engagement for Safe Testing Initiative (AV TEST Initiative) in June 2020 to provide the public with information about the testing of automated driving system-equipped vehicles, in order to promote the safe development, testing, integration, and education of driving automation technology,[19] and expanded it from a pilot to a full program in January 2021. Up to date, NHTSA has not established performance measures and testing standards for automated vehicles and individual manufacturers set their own testing requirements. This results in an inconsistent level of testing on public roads, posing potential safety hazards that remain unverified.[20]

As revealed by the National Transportation Safety Board's investigation,[21] the absence of federal safety risk management requirements for testing automated vehicles on public roads by the NHTSA has led to a situation where, despite some automation developers voluntarily submitting safety self-assessment reports to the NHTSA, these reports remain unevaluated by the agency. This has resulted in a lack of meaningful safety or technical information. In response to this deficiency, certain states have implemented their own risk management-focused requirements that developers must comply with before conducting tests on public roads. The National Transportation Safety Board (NTSB) in its Tempe investigation report recommended that the automation developer (Uber ATG, which has since

[17] Connected & Automated Mobility 2025: Realising the benefits of self-driving vehicles in the UK, 2022 August., https://assets.publishing.service.gov.uk/media/62ff438c8fa8f504cdec92df/cam-2025-realising-benefits-self-driving-vehicles.pdf

[18] See the Automated Vehicles (AV) Bill, https://bills.parliament.uk/bills/3506

[19] https://www.nhtsa.gov/automated-vehicle-test-tracking-tool

[20] https://www.ntsb.gov/Advocacy/safety-topics/Pages/automated-vehicles.aspx

[21] https://www.ntsb.gov/Advocacy/safety-topics/Pages/automated-vehicles-investigative-outcomes.aspx

been acquired by another developer) complete the implementation of a safety management system which would have addressed all the issues that presented a safety risk during the testing of automated vehicles on public roads. NHTSA was also recommended to require automation developers to submit a safety self-assessment report prior to testing and evaluate the reports to determine whether they include appropriate safeguards for testing automated vehicles on public roads.[22]

In June 2021, the NHTSA issued a **Standing General Order on Incident Reporting for Automated Driving Systems (ADS) and Level 2 Advanced Driver Assistance Systems (ADAS)**.[23] This mandate requires manufacturers and operators of vehicles equipped with SAE Level 2 advanced driver assistance systems or SAE Levels 3–5 automated driving systems to report crashes. As currently automated driving systems are still primarily in development or in limited use on public use and not available for sale to the general public, this order mainly pertained to testing vehicles on a large scale. The data collected will aid the agency in identifying potential safety issues and impacts stemming from the deployment of advanced technologies on public roads, thus enhancing transparency. Access to automated driving systems data could reveal common patterns in driverless vehicle crashes or systematic operational issues. NHTSA's oversight extends beyond specified crashes outlined in the order or data submitted under reporting obligations, encompassing all relevant information and incidents pertaining to potential safety defects. Moreover, NHTSA may take additional actions on individual crashes, including deploying Special Crash Investigations teams and requesting further information from the involved companies. If necessary, NHTSA may initiate defect investigations.[24]

China has adopted a stricter regulatory approach towards automated driving road testing, issuing a series of more detailed and stringent regulations on automated driving road testing from the national level down to the local level. Characteristic of China's approach is the introduction of an intermediate phase of demonstration deployment prior to commercial

[22] https://www.ntsb.gov/Advocacy/safety-topics/Pages/automated-vehicles-investigative-outcomes.aspx

[23] NHTSA, Second Amended Standing General Order Incident Reporting for Automated Driving Systems (ADS) and Level 2 Advanced Driver Assistance Systems (ADAS), https://www.nhtsa.gov/sites/nhtsa.gov/files/2023-04/Second-Amended-SGO-2021-01_2023-04-05_2.pdf

[24] https://www.nhtsa.gov/press-releases/nhtsa-orders-crash-reporting-vehicles-equipped-advanced-driver-assistance-systems

deployment in the regulatory framework of roadtesting. This phase serves as an additional validation stage for vehicle functions following road testing. Compared to prescribed verification schemes regarding path, time, and test content, demonstration applications can be closely aligned with societal needs, offering services and executing driving tasks pertinent to public transportation and cargo logistics. Such demonstrations validate vehicle operational capabilities within specific areas and engage in human-computer interaction in real-world settings. Importantly, public demonstration deployments not only verify the effectiveness of autonomous driving technology but also enhance public awareness and trust, laying the groundwork for the forthcoming large-scale commercial deployment of automated driving functions.[25]

Road Testing of Automated Vehicles in China

In April 2018, China's Ministry of Transport, the Ministry of Industry and Information Technology, and the Ministry of Public Security jointly issued the **"Road Test Management Specification for Intelligent Connected Vehicles (Trial),"** with local governments subsequently issuing detailed rules for implementation. These regulations outline conditions and obligations for entities involved in automated driving road tests, including test subjects, drivers, and vehicles, as well as procedures for managing accident data and processing traffic accidents. Automated vehicle road tests are categorized into three types: general technology tests, special technology tests, and trial operation tests.[26] It was soon revised in 2021 and supplemented by the **"Management Specification for Road Test and Demonstration Operation for Intelligent Connected Vehicles (Trial),"** authorizing pilot operation of L3 and L4 automated vehicles in certain areas.", demonstrating the increasingly clear path from road testing to demonstration applications and ultimately to commercial applications for automated driving in China. Up to date, more than 50 provinces or municipalities (with several regions utilizing local legislation or establishing policy pilot zones) all across the country have issued detailed

[25] See Management Code for Road Testing and Demonstration Application of Intelligent Connected Vehicles (Trial) (Draft for Comments), http://www.gov.cn/zhengce/zhengceku/2021-08/03/content_5629199.html

[26] See the Policy Interpretation of The Implementation Rules of Road Test Management of Automated vehicles in Beijing (Trial), http://jtw.beijing.gov.cn/xxgk/zcjd/202011/t20201119_2140422.html

rules for road testing, promoting innovative exploration in manned or unmanned testing, freight testing, high-speed testing, commercial pilot testing, and other testing demonstrations.

Commercial Deployment in China

In order to promote the commercial deployment and mass production of automated vehicles, The Ministry of Industry and Information Technology issued in 2021 "**Access Management Guideline for Intelligent Connected Vehicles**" guiding the automated vehicles enterprise management, product design process and tests as regard to functional safety, Safety of the Intended Functionality, cybersecurity, software update, data security, etc. By the end of 2023, the automated driving industry celebrated two significant milestones on the path to commercial deployment: the "**Notice on Carrying out Pilot Work on Access and Road Access of Intelligent Connected Vehicles**" and the "**Guidelines for Transport Safety Services of Autonomous Driving Vehicles (Trial)**". Especially the last announcement marked the first official recognition, at the national policy level, of intelligent connected vehicles for use in transportation and operational activities. Almost simultaneously, the Ministry of Transport issued the **Automated Driving Vehicle Transport Safety Service Guidelines (Trial)** to guide the development of automated driving technology and regulate the deployment of automated driving vehicles in the transportation service sector.

In addition to ministerial regulations, China has been expediting both local and national legislation to establish a legal framework for automated driving since 2021. In March 2021, the Ministry of Public Security issued the Revised Draft of the Road Traffic Safety Law, delineating regulations for road testing and deployment of vehicles with automated driving features, as well as liability issues.[27] In July 2022, Shenzhen announced the **Shenzhen Special Economic Zone's Intelligent Connected Vehicles Management Regulation**, which comprehensively address road testing and demonstration deployment, access and registration procedures, network security and data protection, collaborative road infrastructure, traffic accident management, and legal liabilities. Notably, this represents China's inaugural local legislation specifically tailored to the commercial deployment of automated driving technology.

[27] https://www.mps.gov.cn/n2254536/n4904355/c7787881/content.html

From closed-course testing to road testing, and from pilot demonstrations to commercial deployment trials, China's automated driving-related industries and markets are currently experiencing rapid growth. Data shows that China has established 17 national-level testing demonstration zones, 7 national-level Internet of Vehicles pilot zones, and 16 "Dual-Intelligence" pilot cities. These initiatives have opened up over 22,000 kilometers of testing roads and accumulated over 70 million kilometers of testing mileage.[28]

Adapting to Dynamic Risk Landscapes in Automated Driving: Regulatory Challenges and Strategies

Autonomy Fallacy and the Legal Status of Automated Driving System

SAE Level 3 automated systems are capable of executing certain driving tasks and monitoring the driving environment under specific conditions. However, human intervention is required to take back control when requested by the automated system. SAE Level 4 automated systems are capable of both conducting driving tasks and monitoring the driving environment without the need for human intervention. However, these systems can only operate within defined environments and under predetermined conditions. SAE Level 5 automated systems possess the capability to perform all driving tasks under any conceivable conditions, equivalent to those that a human driver could execute.[29] As driving operations gradually transition control from human drivers to vehicles, specifically automated driving systems, the legal framework must adapt to address these changes effectively. An essential initial question for every potential traffic accident is, "Who should be held liable?" In a typical vehicle collision, the plaintiff often holds the driver and/or the vehicle manufacturer responsible. However, in cases involving automated vehicles, liability inquiry become exceedingly intricated.

[28] Ye Haoming &Wang Yuhao, Promoting Technological Progress, Stabilizing Market Development, and Ensuring Safety Bottom Line—Insights into the Current Development Standards of Autonomous Driving Technology, China Xinhua News Agency, https://baijiahao.baidu.com/s?id=1784459364749429642&wfr=spider&for=pc

[29] See "Federal Automated Vehicles Policy" https://www.transportation.gov/AV/federal-automated-vehicles-policy-september-2016

As noted by German scholars *Gleis and Jener* (Vgl. Gleis & Jener, 2016, p. 561), In the past century, the automobile, often referred to as a "carriage without horses," has significantly improved our mobility and transformed our road infrastructure. Now, we are entering an era of the "carriage without a coachman," marked by new versions of vehicles and accompanied by fresh risk scenarios. In this evolving traffic landscape shaped by the technological advancements of driving automation, the investigation into liability may reveal a new dimension.

The transformative impact of recent technological advancements in robotics, especially the emergence of robots or AI system capable of learning from experience and take independent decisions sparks the discussion of whether robots should be endowed with legal personhood, which was highlighted by European Parliament Resolution of 16 February 2017 with Recommendations to the Commission on Civil Law Rules on Robotics.[30] Recognizing "the more autonomous robots are, the less they can be considered to be simple tools in the hands of other actors", which prompts questions "whether the ordinary rules on liability are sufficient or whether it calls for new principles and rules to provide clarity on the legal liability of various actors concerning responsibility for the acts and omissions of robots where the cause cannot be traced back to a specific human actor and whether the acts or omissions of robots which have caused harm could have been avoided", the resolution contends "the autonomy of robots raises the question of their nature in the light of the existing legal categories or whether a new category should be created, with its own specific features and implications".

The Resolution recommend to consider the legal solution by "creating a specific legal status for robots in the long run, so that at least the most sophisticated autonomous robots could be established as having the status of electronic persons responsible for making good any damage they may cause, and possibly applying electronic personality to cases where robots make autonomous decisions or otherwise interact with third parties independently". Interestingly, while attempting to define the autonomy of the intelligent robot as "the ability to take decisions and implement them in the outside world, independently of external control or influence", the resolution admits that this autonomy is purely technical in nature, and its

[30] See "European Parliament Resolution of 16 February 2017 with Recommendations to the Commission on Civil Law Rules on Robotics (2015 /2103 (INL)). http://www.europarl.europa.eu/doceo/document/A-8-2017-0005_EN.html? redirect

extent is determined by the sophistication level with which the robot is designed to engage with its surroundings.

Apart from its feasibility of implementation, this proposal exhibits logical inconsistency and legal ambiguity, partly due to the interchangeability and confusion between the terms "autonomous" and "autonomy." Autonomy, often attributed to Kant's philosophy, means the ability of a rational being's will to act as a law unto itself, free from the influence of any external factors or properties of objects of desire. Autonomy encompasses the capacity for self-governance and self-determination, allowing individuals to make rational assessments and decisions based on facts and reasons. This involves not only possessing the capability for self-governance but also effectively exercising it by scrutinizing one's values, engaging in self-determination, and adhering to moral principles. Furthermore, autonomy is shaped by social and historical contexts as individuals develop their values within their communities (Rubel et al., 2021, p. 24). Accordingly, autonomy is not an ontological concept identical to automation, but rather a normative construction. Degrading autonomy to a purely technical level will diminish its normative essence.

The concept of technical autonomy can be misleading despite of the sophistication of the machine learning system. Machine learning algorithms represented by neural network algorithm are increasingly applied in automated vehicle's decision-making, for instance, Bojarski, based on the high-performance GPU and the parallel computing tool CUDA, used convolutional neural network (CNN) to train the front camera and build an end-to-end decision system that directly output steering wheel angles, skipping over the human decomposition behavior and indicator formulation in the intermediate processes, thus to improve the performance of the decision system (A Review of Academic Research in Chinese Automotive Engineering 2017). But we still should keep in mind that it is quite different from human's psychological and social cultural decision-making mechanism. Although automated driving systems are designed to be able to process the internal vehicle operation data and the external environment data collected by sensors through highly complex computing to make decisions and activates the corresponding automatic operation, it may be hard to regard this process as the manifestation of a free will or an act in the civil law's sense. Programmed by human developers, automated driving algorithms operate based on predefined rules or parameters, logic, and instructions set by humans and is not capable of self-governance and self-determination. Additionally, automated driving algorithms are subject to external control and oversight by human operators or administrators. Human intervention remains indispensable in SAE Level 2–4 automated vehicles, as well as in SAE Level 5 vehicles, where input from a human passenger is still required to specify the destination. The reliance on human intervention further highlights the lack of autonomy inherent in automated driving systems. According to the SAE J3016, the term "autonomous," rooted in the realms of robotics and artificial intelligence, signifies systems capable of independent and self-sufficient decision-making. As autonomy in jurisprudence denotes the ability for self-governance, labeling automated driving technology as "autonomous" is inaccurate, given that even the most sophisticated automated driving systems primarily operate based on algorithms and comply with user commands.[31]

As the study report on Artificial Intelligence and Civil Liability Legal Affairs contend, there are no technological, philosophical, or legal justifications to consider machines either equivalent to humans or belonging to an intermediate category between the two. Specifically, it is unreasonable to attribute moral responsibility to a machine, as it lacks the capacity to choose its actions and does not possess intentionality that could be influenced by ex ante incentives, such as legal sanctions. In other words, machines cannot be held accountable or coerced, as they do not fear sanctions, since they do not perceive or value their existence and do not seek to preserve it for their own sake.[32]It is therefore not surprising when in December 2018, the European Commission-appointed High-Level Artificial Intelligence Expert Group submitted draft ethical guidelines for trustworthy AI which abandoned the approach of the European Parliament's "Legislative Recommendation to the European Union on Civil Law Rules on Robotics" of entitling AI with electronic personhood, arguing that AI has no self-purpose in term of its development and use and it is merely a tool to enhance human well-being and is therefore not a subject of law, but an object of law (Vgl. Dettling & Krueger, 2019).

With the legal status of the automated system being denied, the question of liability for damages, injuries, or fatalities resulting from the operation of automated driving systems remains unanswered and requires

[31] See Taxonomy and Definitions for Terms Related to Driving Automation Systems for On-Road Motor Vehicles J3016_202104, https://www.sae.org/standards/content/j3016_202104/

[32] Artificial Intelligence and Civil Liability Legal Affairs noted, study requested by the JURI committee requested by the European Parliament's Committee on Legal Affairs, https://www.europarl.europa.eu/RegData/etudes/STUD/2020/621926/IPOL_STU(2020)621926_EN.pdf.

further exploration. This inquiry is highly contentious due to the intricate interaction of various technologies employed in perception, artificial intelligence, and control systems, particularly when considering the Human-Machine-Interface problem.

Human-Machine-Interface (HMI), Control Problem and Liability Gap

The so-called control problem arising from the tendency of the human agent within a human–machine control loop to become complacent, over-reliant or unduly diffident when faced with the outputs of a reliable autonomous system, has been recognized for many years by industrial psychologists and engineers studying the human operators of complex machines (Zerilli et al., 2019). The convergence of emerging technologies and their inherent risks makes the human-machine interaction in automated driving potentially more hazardous than in other Internet of Things (IoT) applications. As autonomy levels advance from Level 1 to Level 5, automated driving systems assume a progressively larger share of driving tasks, gradually reducing human involvement until it is completely eliminated at Level 5. Consequently, human-machine interaction issues are most prominent at Levels 3 and 4. High levels of automation can lead to "out-of-the-loop" issues like complacency, skill degradation, mental underload (when automation functions reliably), mental overload (when operators must suddenly address automation-induced problems), and reduced situation awareness (Lu et al., 2016).

1. The Technological Description of HMI in SAE Levels 3–4 Automated Driving as defined by SAE J3016

SAE Level 3 mandates that the human driver remain alert while activating the automated driving mode to regain control when prompted by the system or in the event of system failure. However, the limited capacity of humans to maintain alertness when disengaged from driving tasks poses a significant safety concern. Human nature tends to lead people to over-trust technology once they witness it functioning correctly. Consequently, individuals may psychologically disengage from the operation of the technology, making it difficult to intervene just in time when necessary. It is a typical phenomenon of responsibility split in the management area or automation bias in the engineer field (Lipson & Kuerman, 2017, pp. 72–73). This tendency can be particularly perilous in high-speed operating machines like automobiles. A study conducted by the Virginia Tech Transportation Institute revealed that the average response time for Level 3 automated vehicle drivers to take over control was 17 seconds, during which a vehicle traveling at a speed of 105 km/h had covered a distance of 494 meters (Liu et al., 2017, p. 7). Additional research also indicates a substantial delay, ranging from 8 to 40 seconds, in an individual's ability to reorient themselves to the driving task (Agrawal et al., 2017; Merat et al., 2014). When a crash occurs during the take-over process, determining the party at fault can be challenging. Filling the liability gap will require scrutinizing both the human driver's reaction and the Human-Machine Interface (HMI) design concerning how the vehicle conveys alert signals or other safety-related information to the human driver, operator, occupants, and even external actors.[33]

SAE level 4 allows the vehicle to operate completely autonomously under certain conditions, such as in a defined route or in certain traffic scenes. The system will fall back to the minimal risk condition rather than rely on human intervention in case of system incompetency or failure. Unlike at Level 3, the Level 4 feature user is "……not expected to respond to a request to intervene in order to perform the fallback. Nevertheless, in the case that a Level 4 sub-trip feature reaches its operational design domain (ODD) limit, the automated driving systems may issue an alert to the passenger that s/he should resume driving in order to complete their trip".[34] As the liability gap between the human user (or driver) and automated driving systems under Level 4 appears to narrow, the direct interaction in the form of fallback between human and automated driving systems is replaced by the requirement for the automated driving systems to maintain a minimal-risk condition. However, uncertainty persists regarding who should be held accountable if the minimal-risk condition maneuver fails to prevent a crash and the human onboard fails to respond in time.

A study on the control problem of HMI suggests that higher levels of automation could potentially exacerbate control difficulties, thus impacting the safety of the HMI system in a negative manner (Zerilli et al., 2019). In the context of vehicle automation, such as in SAE Level 1, there is a clear delineation of tasks between drivers and vehicle subsystems.

[33] Federal Automated Vehicles Policy 2016, P.22.

[34] Taxonomy and Definitions for Terms Related to Driving Automation Systems for On-Road Motor https://www.sae.org/standards/content/j3016_202104/

However, as automation levels increase, this allocation becomes blurred, leading drivers to struggle in accurately assessing the vehicle's capabilities and often overestimating them. Thus, paradoxically, while larger and more sophisticated technologies are generally reliable, they may pose greater control issues than smaller, simpler ones. Indeed, as automated driving technology evolves and SAE J3016 undergoes revisions, the HMI challenge may also worsen.

Notably, the 2021 version introduced significant changes in HMI issues, incorporating new terminologies to accommodate remote technologies and designate individuals involved in remotely controlling a vehicle: "remote assistant" and "remote driver." According to the SAE J3016 standard, remote assistance refers to the event-driven provision of information or advice by a human located remotely to an automated driving system-equipped vehicle operating in driverless mode. The purpose of remote assistance is to aid the vehicle in continuing its trip when it encounters a situation that it cannot manage autonomously (SAE J3016 3.15). Remote driving, as defined by the SAE J3016 standard, involves the real-time control of a vehicle's motion by a human driver located remotely from the vehicle. The remote driver performs actions typically associated with driving, such as braking and steering, and has the authority to make decisions regarding the vehicle's motion, overriding the automated driving systems when necessary (SAE J3016 3.24). While both remote assistance and remote driving involve human intervention in the operation of an ADS-equipped vehicle, they differ in terms of the level of control exerted by the human operator, the purpose of their involvement, and the extent of their authority over vehicle motion. As indicated by report, SAE distinguish between remote assistance and remote driving in that with assistance the automated driving systems rather than the human carry out the maneuver, the automated driving systems must assess whether the instructions received from the remote assistant represent a safe path ahead and also whether the maneuver is legal, for a vehicle to be truly self-driving.[35] Even though a remote assistant is not driving, they might still cause harm. For instance, as illustrated by the SAE, a remote assistant could mistakenly misidentify an obstacle as a plastic bag when it actually contains concrete.[36]

[35] Automated Vehicles: joint report, Law Commission, https://www.lawcom.gov.uk/project/automated-vehicles/, pp. 167.

[36] Automated Vehicles: joint report, Law Commission, https://www.lawcom.gov.uk/project/automated-vehicles/, pp. 167.

However, achieving a clear and definitive distinction between these two concepts is not always feasible. For example, when a human remotely steers a vehicle through a construction zone, it qualifies as remote driving. Conversely, providing a travel path for the vehicle to follow using its own sensors is categorized as remote assistance, falling outside the scope of automated driving function according to SAE J3016 standard. The line between automated/remote driving and remote assistance may blur in scenarios necessitating a blend of automated driving capabilities and direct human driving maneuvers, for example a remotely driven vehicle might integrate automated features to maintain safe operation if its connection to the remote human is disrupted (Smith, 2022).

The introduction of remote assistance and remote driving could exacerbate the HMI issue, necessitating stricter requirements for the automated driving systems regarding instantaneous and comprehensible communication mechanisms. At the same time, it poses more stringent requirements regarding the situation awareness and responsiveness of the human agents providing remote assistance or remote driving. This challenge may be more pronounced under SAE Level 4, where the automated driving system becomes more self-sufficient and less reliant on human intervention (yet still may require remote assistance or remote driving), as the conundrum of control in HMI arises from the fact that as a system becomes more reliable, it becomes increasingly challenging for a human supervisor to stay sufficiently engaged with the technology to ensure a safe transition to manual control in case of malfunction (Zerilli et al., 2019).

2. Legislative and Regulatory Capture of HMI Issue

German Eighth Amendment to the Road Traffic Act of 2017[37] addressed the human-machine co-driving issue by defining the highly or fully automated driving vehicles and their human driver's rights and obligations. According to Article 1a (2) this amendment, the automated driving systems should be capable of conducting driving tasks, in compliance with traffic regulations, recognizing the necessity of human control and alerting the driver through sensory means. The human drivers when activating automated driving systems should remain vigilant and be ready to regain control when requested by the system or "when the driver is supposed to recognize or is supposed to recognize based on obvious

[37] https://dserver.bundestag.de/btd/18/085/1808559.pdf

circumstances that the highly or fully automated driving systems no longer meets the functional conditions for use" (Article 1b). However, there are concerns raised by scholars and the public regarding the practicality and compatibility of these regulations with advanced automated driving technology, as they may impose overly strict obligations on drivers and fail to align with the intended development of automated systems (Vgl. Koenig, 2017). Additionally, ambiguity persists regarding the timing and manner of transferring control between humans and machines, further complicating the implementation of effective human-machine interactions in automated driving contexts.

The German legislation's pioneering approach to legalizing automated driving on a national level seems to have impacted Chinese legislators and regulators in defining the HMI (Human-Machine Interface) concerning the vigilance and immediate take-over obligation of the human drivers. According to the Chinese **"Management Specification for Road Test and Demonstration Operation for Intelligent Connected Vehicles (Trial)**, during road tests and demonstration deployments, drivers must continuously monitor the vehicle's operational status and the surrounding environment. If they notice that the vehicle is not suitable for automated driving or if the system prompts manual intervention, they must promptly take appropriate action (Article 26). **Road Traffic Safety Law (Revised Draft) of the Ministry of Public Security** also stipulates the obligation of the driver to be alert and the obligation to take over the vehicle (Article 155). Whether the above provisions are appropriate is worth pondering: First, it is not easy to determine the scope of the driving behavior and the extent of driving behavior constraints. **Shenzhen Special Economic Zone's Intelligent Connected Vehicles Management Regulation** stipulates that when operating in the automated mode of conditional and high-level automated driving, drivers are required to remain in the driver's seat, actively monitoring both the vehicle's operations and its surroundings, and be prepared to take control at a moment's notice. Should the vehicle request a takeover or encounter conditions unsuitable for automated driving, the driver must promptly assume manual control (Article 35).

Establishing precise behavioral norms regarding HIM under L3-L4 automated driving technology can be very challenging. German and Chinese legislator's approach of imposing continuous vigilance and immediate takeover obligations contradicts the technical capabilities of automated driving systems. Not monitoring the environment while using automated driving does not inherently violate the duty of care; it only violates the duty when minimum vigilance requirements are breached: The driver who activate automated driving functions should ensure that he or she can immediately take control of the vehicle if necessary. However, defining the duty of minimum vigilance is also very challenging and essentially leads to imposes a general duty of care (Lüdemann, 2018, p. 414). The appropriate approach is to transfer the obligation of environmental monitoring to vehicles based on the characteristics of L3 and above automated driving technology and abolish provisions regarding vigilance obligations. When vehicles are operating in automated driving modes above L3, the system is obliged to monitor the environment within the operational range and promptly notify human drivers of unexpected traffic situations. Drivers are obligated to take over the vehicle upon receiving notification. If a driver fails to take over in a timely manner, resulting in a traffic accident, the driver bears responsibility for the traffic negligence. If the vehicle fails to issue notification information or the notification is untimely, the manufacturer should assume the product liability for product defects.

The recent revised German Road Traffic Law by **the Autonomous Driving Act in 2021**[38] allows autonomous vehicle operation without a physically present driver but requires “technical supervision” in designated areas, aligning with SAE-Level 4 standards. The updated legal framework mandates autonomous vehicles to maintain continuous radio connectivity with a designated technical supervisor, responsible for remote deactivation and driving maneuvers, termed “technical supervision.” The automated driving systems shall promptly notify the technical supervisor when the continuation of the journey would require a violation of road traffic regulations and propose potential driving maneuvers to the technical supervisor, providing necessary data for assessing the situation. The technical supervisor is obligated to evaluate the alternative driving maneuvers indicated by the vehicle system and approve them if they do not compromise traffic safety. This approval or resolution shall then be reviewed and executed by the automated driving system if it is deemed safe. If not, the automated driving system will decide to enter the minimum risk condition, in which the technical supervisor shall immediately establish contact with the occupants of the motor vehicle and initiate the necessary measures to ensure traffic safety. The German legislature introduces the

[38] https://www.bundestag.de/dokumente/textarchiv/2021/kw20-de-autonomes-fahren-840196

position of "technical supervisor" as a human operator tasked with both remote assistance and remote driving duties, aiming to maintain human oversight and ensure driving safety.

The Chinese **Automated Driving Vehicle Transport Safety Service Guidelines (Trial)** also introduces a remote safety operator for highly automated vehicles. It stipulates that: "… for fully automated taxis, with approval from the municipal government, remote safety operators can be used in designated areas, with a minimum ratio of one safety operator per three vehicles. Safety operators must undergo training in automated driving technology, transportation regulations, and emergency handling skills. They should also stay updated on changes or upgrades in automated driving systems. Additionally, they must meet relevant qualifications and obtain appropriate certifications for their roles in transportation" (Art.6).

While the legislature's intention to integrate both remote assistance and remote driving capabilities for human oversight is aimed at effectively managing unforeseen circumstances, the underlying HMI challenge may ultimately hinder the realization of this ambition. Setting aside the debate on whether a redundant check and balance mechanism between the human agent (technical supervisor) and the automated driving systems is desirable for achieving safety goals, the inclusion of "technical supervision" in German road traffic law underscores the HMI challenge within the SAE Level 4 standard. On one hand, it demands seamless integration of human and machine factors, along with reliable data transmission; on the other hand, it introduces more challenging context related to situation awareness and the decision-making process. Situation awareness refers to the ability to perceive, comprehend, and anticipate elements of the environment relevant to achieving a goal, such as safely navigating a vehicle. In the context of automated driving, situation awareness involves monitoring various factors, including road conditions, traffic patterns, weather, and the vehicle's performance. Technical supervisors encounter a remote or virtual situation to be aware of: they need to monitor real-time data streams from automated vehicles to identify potential risks or anomalies. This might require continuous attention to system alerts, notifications, and sensor readings to detect deviations from normal operation. However, as discussed above, as ADSs become increasingly reliable and human intervention is expected to occur less frequently, maintaining situational awareness becomes more challenging due to the potential pitfalls of overtrust and complacency.

Overall, the provision of technical supervision within the German road traffic law reflects the intricate interplay between human agent and the ADS. As automated vehicle technology continues to evolve, addressing these HMI challenges become even more crucial in realizing the full potential of automated driving while maintaining safety and efficiency on the roads.

The recently announced UK AV Bill introduces a new legislative approach that offers a more realistic view of Human-Machine Interface (HMI) challenges by allowing a "transition period." It defines "transition demand" as "a demand, communicated by equipment of a vehicle in which an authorised user-in-charge feature is engaged, that the user-in-charge assume control of the vehicle by the end of a period of time beginning with the communication of the demand (the 'transition period')". According to this article, the Secretary of State must impose authorisation requirements to ensure that"(a) the transition demand will be capable of being perceived by anyone who might legally be a user-in-charge of the vehicle (having regard in particular to users-in-charge with disabilities), (b) the transition period will be long enough for the user-in-charge to prepare to assume, and assume, control of the vehicle, (c) the vehicle will continue to travel autonomously, safely and legally during the transition period, (d) equipment of the vehicle will make a further communication at the end of the transition period to alert the user-in-charge to the ending of the period, (e) and the vehicle will deal safely with a situation where the user-in-charge fails to assume control by the end of the transition period." (Section 7 (2)).

The requirement for a sufficient transition period for users to prepare and assume control acknowledges the potential cognitive and psychological challenges associated with HMI during the transition from automated to manual driving. During this period, the vehicle is supposed to continue operating reliably, reflecting the minimum risk condition as stipulated by the SAE J3016 Standard. It can be expected that by requiring the automated driving systems to communicate the transition demand and provide additional alerts at the end of the transition period, the provision enhances both the situational awareness and responsiveness of the human agent. The liability immunity article for the user in charge states that if an offense occurs while the individual was the user-in-charge of the vehicle, and if during that time, the individual's driving did not fall below the standard expected of a careful and competent driver given the circumstances, the bill also exempts the user in charge from liability for the conduct during the transition period (Sections 47–48). By transferring the HMI risk from the human agent to the AV system, this approach imposes a higher duty of care standard for HMI design on the system developer. However, the requirement for vehicles to safely manage such scenarios demands sophisticated and highly reliable autonomous systems, which on one hand can be challenging to develop, and on the other hand, may inadvertently encourage over-reliance on the machine.

Conclusion

Regulating automated driving poses a formidable challenge, if not an impossible one. The continuous evolution of related technologies introduces uncertainty regarding their solution, mitigation, or potential escalation of those risks. In broad terms, regulating emerging technologies poses a challenge marked by uncertainty, alongside the aspiration to keep up to date with rapidly evolving technological progress. "Contemporary scientific knowledge is characterized increasingly by uncertainty. This is due not only because both the risks and the unpredictability linked to it are increasing, but above all because of the intrinsic incompleteness and indeterminacy of scientific knowledge compared with the needs to make social choices, public policy, and legal decisions." (Tallacchini, 2005, p. 647) In conclusion, navigating the risk landscape of AV regulation necessitates a thorough comprehension of the risks and associated uncertainties inherent in this complex domain. Indeed, understanding these risks is integral to the regulatory process itself. The various existing regulatory approaches, whether successful or unsuccessful, serve as invaluable examples for calibrating uncertainty and refining future regulatory strategies. By learning from these experiences, policymakers and stakeholders can work towards developing more effective and adaptive regulatory frameworks that address the evolving challenges of automated driving technology.

References

A Review of Academic Research in Chinese Automotive Engineering. (2017). *Chinese Highway Journal, 6.*

Agrawal, R., et al. (2017). Effects of change in environment on the minimum time to situation awareness in transfer of control scenarios. *Transfer of Control Scenarios Transportation Research Record, 2663*(1), 126–133.

Gasser, T., & Westhoff, D. (2012). *BASt-study: Definitions of automation and legal issues in Germany*. TRB Road Automation Workshop. http://onlinepubs.trb.org/onlinepubs/conferences/2012/Automation/presentations/Gasser.pdf

Hilgendorf, E. (2016). *Automatic driving and law* (H. Xiaoyan, Trans.). In J. Yi 易继明 (Ed.), *Private law* (私法), Law Press.

Li, Li 李力 & Wang, Feiyue 王飞跃. (2016). *Intelligent vehicles: Advanced sensing and control* (智能汽车:先进传感与控制). Machinery Industry Press.

Li, Keqiang 李克强 et al. (2017). Development status and trends of intelligent connected vehicles technology (智能网联汽车(ICV)技术的发展现状及趋势). *Journal of Automotive Safety and Energy* Efficiency, 1, 1–14.

Lipson, W., & Kuerman, M. (2017). *Driverless* Driving (L. Luyin & J. Yang, Trans.). Wenhui Press.

Liu, S., Jie, T., Shuang, W., & Liyun, L. (2017). *The first driverless technology book*. Electronic Industry Press.

Lu, Z., et al. (2016). Human factors of transitions in automated driving: A general framework and literature survey. *Transportation Research Part F: Traffic Psychology and Behaviour, 43*, 183–198.

Lüdemann, S. (2018). Vogelpohl: Neue Pflichten für Fahrzeugführer beim automatisierten Fahren–eine Analyse aus rechtlicher und verkehrspsychologischer Sicht (NZV 411,414).

Merat, N., Jamson, H., Lai, F., Daly, M., & Carsten, O. (2014). Transition to manual: Driver behaviour when resuming control from a highly automated vehicle. *Transportation Research Part F: Traffic Psychology and Behaviour, 27*, 274–282.

Rubel, A., Castro, C., & Pham, A. (2021). *Algorithms and autonomy: The ethics of automated decision systems*. Cambridge University Press.

Smith, B. W. (2022). On remote driving., https://cyberlaw.stanford.edu/blog/2022/05/remote-driving.

Tallacchini, M. (2005). Before and beyond the precautionary principle: Epistemology of uncertainty in science and law. *Toxicology and Applied Pharmacology, 207*(2), 645–651.

Vgl. Dettling & Krueger. (2019). Erste Schritte im Recht der Künstlichen Intelligenz, MMR, 211,212.

Vgl. Gleis & Jener. (2016). Hochautomatisiertes und autonomes Autofahren-Risiko und rechtliche Verantwort -ung，JR 561.

Vgl. Koenig & Die. (2017). gesetzlichen Neuregelungen zum automatisierten Fahren, NZV 123, 124., m. w. N.

Zerilli, J., et al. (2019). Algorithmic decision-making and the control problem. *Minds & Machines, 29*, 555–578.

CHAPTER 7

Exploring a Suitable Legal Framework of Civil Liability for Automated Vehicle Accidents

Chunyan Ding

Introduction

According to the World Health Organization's report, about 1.3 million people lose their lives, and between 20 and 50 million people suffer personal injuries in road traffic accidents every year.[1] Road traffic crashes cost most countries three per cent of their gross domestic product and become the leading cause of death for children and young adults aged between five to twenty-nine years.[2] The data from the Ministry of Public Security show that in mainland China, there were 453 million automobile cars and

[1] World Health Organization, 'Road Traffic Injuries', https://www.who.int/news-room/fact-sheets/detail/road-traffic-injuries accessed 19 Feb 2024.

[2] Ibid.

C. Ding (✉)
School of Law, City University of Hong Kong, Kowloon, Hong Kong
e-mail: chunding@cityu.edu.hk

C. Ding, R. Fan (eds.), *Chinese Ethical and Legal Perspectives on Automated Vehicles*,
https://doi.org/10.1007/978-981-96-1277-2_7

542 million drivers in 2024. In 2018, there were 244,937 road traffic accidents nationwide that resulted in a death toll of 63,194 persons and direct financial losses of 1.385 trillion yuan.[3]

Human error in driving (such as speeding, drunk or drugged driving, distracted driving, etc.) is the primary reason for car accidents. European Union's data showed that ninety per cent of car accidents were caused by human error.[4] It is critical to ensure safe vehicles driven by safe drivers on safe roads. Hence, vehicles that apply the evolving automotive technology aim to enhance road traffic safety by reducing or eliminating human error in driving so that significant personal, financial, and societal costs related to vehicle crashes can be saved. Currently, vehicles equipped with automated driving system features are running on the road. Highly automated vehicles as future transport are expected to increase road safety and mobility for elderly and disabled people, cut environmental costs, and improve efficiency and convenience (Carp, 2018).

Therefore, the Chinese government firmly encourages the development of automated vehicles. In September 2019, the Central Committee of the Chinese Communist Party and the State Council jointly issued the *Outline for the Construction of Nation with Strong Transportation System*.[5] The enactment states in Article 3(1) that "the research and development of intelligent networked vehicles (intelligent vehicles, automated, and cooperative vehicle infrastructure system) shall be strengthened to create an independent, controllable, and complete industry chain." In February 2020, the National Development and Reform Commission led ten other governmental departments to release the *Innovative Development Strategy for Intelligent Vehicles*.[6] In December 2020, the Ministry of Transport announced the *Guiding Opinions on Promoting the Development and*

[3] 'One Fatal Car Accident Every Eight Minutes' (每8分钟就有1人死于车祸), *Xinhua News*, 2 Dec 2020, http://www.xinhuanet.com/politics/2020-12/02/c_1126809938.htm accessed 19 Feb 2024.

[4] 'Road Safety Facts', https://roadsafetyfacts.eu/road-safety-what-progress-has-been-made/, accessed 19 Feb 2024.

[5] The *Outline for the Construction of Nation with Strong Transportation System* (交通强国建设纲要), http://www.gov.cn/zhengce/2019-09/19/content_5431432.htm accessed 19 Feb 2024.

[6] The *Innovative Development Strategy for Intelligent Vehicles* (智能汽车创新发展战略), https://www.ndrc.gov.cn/xxgk/zcfb/tz/202002/P020200224573058971435.pdf accessed 19 Feb 2024.

Application of Automated Technologies in Road Traffic.[7] The government set 2025 as the target to achieve a basic shape of its smart vehicle technology innovation, industry ecosystem, infrastructure, regulations and standards, and product supervision and data security systems. With the supportive government policy, more and more testing and demonstration areas around the country for road testing and practical demonstrations of automated cars have been established. Up to 2021, there were more than twenty testing and demonstration areas across the nation, and more than 108 companies located in twenty-five Chinese cities have been granted licenses for testing automated vehicles on the open road (Feng, 2021, p. 65).

At the same time, automated vehicles have created heated discussions about ethical and legal challenges, including civil liabilities for traffic accidents they cause. For example, the United States National Highway Traffic Safety Administration data revealed 736 clashes of vehicles on Autopilot since this advanced driver-assistance technology was introduced in 2014, including 17 fatal accidents.[8] The manufacturer Tesla faced investigations, lawsuits and the risk of legal liabilities.[9] It is imminent and essential to explore a suitable legal framework of civil liability for automated vehicle accidents if a government is determined to develop and promote the automated vehicles industry.

Thus, this chapter aims to explore a suitable nationwide legal framework of civil liability for automated vehicle accidents under Chinese law to balance safety enhancement and the promotion of using automated vehicles. It consists of six parts. Part Two provides an overview of China's current national and local regulations on automated vehicles. The following two parts discuss the civil liability issue according to the different levels of automation that automated vehicles have: Part Three focuses on the

[7] The *Guiding Opinions on Promoting the Development and Application of Automated Technologies in Road Traffic* (关于促进道路交通自动驾驶技术发展和应用的指导意见), http://www.gov.cn/zhengce/zhengceku/2020-12/30/content_5575422.htm accessed 19 Feb 2024.

[8] '17 Fatalities, 736 Crashes: The Shocking Toll of Tesla's Autopilot', *Washington Post*, 10 Jun 2023, https://www.washingtonpost.com/technology/2023/06/10/tesla-autopilot-crashes-elon-musk/ accessed 19 Feb 2024.

[9] 'Before Tesla Recalled 2 Million Vehicles, its "Autopilot" Feature Faced Investigations and Lawsuits', *Forbes*, 13 Dec 2023, https://www.forbes.com/sites/zacharyfolk/2023/12/13/before-tesla-recalled-2-million-vehicles-its-autopilot-feature-faced-investigations-and-lawsuits/?sh=751ab0b0292f accessed 19 Feb 2024.

establishment and scope of civil liability of a user-in-charge, and Part Four investigates multiple potential parties who may be required to bear civil liability where a highly automated vehicle causes an accident. Part Five further probes into the liability insurance models that can fit the legal framework of civil liability for automated vehicle accidents and effectively distribute the legal risks. The final part concludes the chapter.

Overview of the Emerging Regulations on Automated Vehicles in China

Likewise, in China, automated vehicles are classified according to the six-level classification of automation: level 0 vehicles involve no driving automation, level 1–2 vehicles apply low-level driving automation and still require human control and supervision, and level 3–5 vehicles ("highly automated vehicles") can autonomously perform all driving tasks in certain environments and road conditions or under all conditions that a human could plausibly carry them out.[10] Specifically, Article 37 of the *Rules for the Administration of the Road Testing and Demonstrative Application of Intelligent Connected Vehicles (for Trial Implementation)*[11] defines three types of highly automated vehicles with different levels of automation:

- "Conditional self-driving" refers to a situation where "all dynamic driving tasks are completed under the operational design condition of the system, and drivers need to provide appropriate intervention in response to the system's dynamic driving task takeover requests" (level 3 automated vehicles).

[10] The Society of Automotive Engineers, Taxonomy of Definitions for Terms Related to Driving Automation Systems for On-Road Motor Vehicles (Revised on 16 June 2018). The Chinese government also adopts the same standards; see the *Taxonomy of Driving Automation for Vehicles* (汽车驾驶自动化分级) issued by the State Administration for Market Regulation in September 2021, https://www.chinabuses.com/uploadfile/2020/0310/20200310123023480.pdf accessed 19 Feb 2024.

[11] The *Rules for the Administration of the Road Testing and Demonstrative Application of Intelligent Connected Vehicles (for Trial Implementation)* (智能网联汽车道路测试与示范应用管理规范(试行)) was jointly issued by Ministry of Industry and Information Technology, Ministry of Public Security, and Ministry of Transport on 27 July 2021, https://www.gov.cn/zhengce/zhengceku/2021-08/03/content_5629199.htm accessed 19 Feb 2024.

- "Highly self-driving" refers to a situation where "all dynamic driving tasks are completed under the operational design condition of the system, the system will request drivers to take over dynamic driving tasks under special circumstances, and drivers/passengers are allowed not to respond to the system requests" (level 4 automated vehicles).
- "Full self-driving" refers to a situation where "the system can complete the dynamic driving tasks in all road environments that drivers can complete, requiring no driver/passenger intervention" (level 5 automated vehicles).

At the national level, no statutory rules have specifically regulated civil liability for automated vehicle accidents. However, the relevant departments of the State Council have promulgated a set of departmental rules concerning rod testing, production licenses, and data protection and security regarding automated vehicles.[12] Moreover, the Department of Public Security issued a draft of the *Amendments to the Road Traffic Safety Law*[13] for public consultation in March 2021. Article 155 of the draft proposes the traffic safety requirements for vehicles with self-driving functions and legal liabilities for traffic accidents caused thereby. Still, the draft provision was not included when the *Road Traffic Safety Law*[14] was amended by the Standing Committee of the National People's Congress in April 2021.

At the local level, local governments (Beijing, Shanghai, Shenzhen, Chongqing, and Zhejiang Province as the pioneers) have been actively exploring their own legal standards and rules for road testing, demonstrative application and operation of automated vehicles. Since the Central Government announced the *Innovative Development Strategy for Intelligent Vehicles* in 2020, more local governments have issued local departmental rules regulating automated cars. Up to now, provincial or municipal governments in twenty-four provincial regions have made local

[12] Such as the *Rules for the Administration of the Road Testing and Demonstrative Application of Intelligent Connected Vehicles (for Trial Implementation), the Opinion on Strengthening the Administration of the Access of Intelligent Connected Vehicles Manufacturers and Products* (关于加强智能网联汽车生产企业及产品准入管理的意见), the *Several Provisions on the Management of Automobile Data Security (for Trial Implementation)* (汽车数据安全管理若干规定(试行)).

[13] The *Amendments to the Road Traffic Safety Law* (道路交通安全法(修订建议稿)), http://www.gov.cn/xinwen/2021-04/03/content_5597682.htm accessed 19 Feb 2024.

[14] The *Road Traffic Safety Law* (道路交通安全法), http://tzb.jiaxing.gov.cn/art/2021/9/13/art_1229203411_58811623.html accessed 19 Feb 2024.

enactments promoting the testing and application of intelligent networked vehicles.[15]

Among them, the Government of Shanghai and the Government of Chongqing issued their local administrative rules with higher legal enforcement than their local departmental rules. For example, Shanghai published the *Measures of Shanghai on the Administration of the Test and Application of Intelligent Connected Vehicles* ("Shanghai Measures")[16] in December 2021. Moreover, the People's Congress of Shenzhen passed the *Regulations of Shenzhen on the Administration of Intelligent Connected Vehicles* ("Shenzhen Regulations")[17]on 23 June 2022; the local law took effect on 1 August 2022. These local enactments mainly consist of regulatory rules for road testing, demonstrative application and operation of automated vehicles, and at the same time address the civil liability of automated vehicles for traffic accidents as well as insurance mechanisms. The relevant local rules at the current legal experiment stage remain fragmented and inconsistent.

Automated Vehicles with a User-in-Charge

Despite an automated driving system, a human user-in-charge remains in an automated vehicle with driving automation (such as level 1–2 automated cars,[18] and level 3–4 automated vehicles when the system occasionally alerts drivers to take over dynamic driving tasks) and may take over driving upon a transition demand or out of a conscious choice. The user-in-charge's role and civil liabilities for car accidents largely share similarities with those of a traditional car driver. This part examines the civil liabilities of a user-in-charge for a car accident under Chinese law.

[15] PKU Law Database (as of 19 Feb 2024).

[16] The *Measures of Shanghai on the Administration of the Test and Application of Intelligent Connected Vehicles* (上海市智能网联汽车测试与应用管理办法), https://www.zjsfq.gov.cn/html/1/169/181/182/246/2109.html accessed 19 Feb 2024.

[17] The *Regulations of Shenzhen on the Administration of Intelligent Connected Vehicles* (深圳经济特区智能网联汽车管理条例), https://finance.sina.com.cn/china/dfjj/2022-07-11/doc-imizmscv0970497.shtml accessed 19 Feb 2024.

[18] Tesla's Autopilot system requires the driver's hands-on and eyes-on attention and, therefore, is classified as a Level 2 driving automation system.

Establishment of Civil Liabilities of a User-in-Charge for Car Accidents

The current legal rules governing the negligence liability of drivers of traditional cars are embodied mainly in two sources of law. One is Chapter Five (titled "Liability for Motor Vehicle Traffic Accident") of Book Seven on Tort Liability of the *Civil Code*.[19] The other is Chapter Five (titled "Handling of Traffic Accidents") of the *Road Traffic Safety Law*. As analyzed below, these relevant legal rules can be applied by analogy for determining the civil liability of a user-in-charge of a low-level automated vehicle.

Above all, in the same way as a driver of traditional vehicles, a user-in-charge of an automated vehicle owes a duty of care to the road users outside the car and the passengers inside the vehicle when he controls it. For example, a user-in-charge is required to be in the driver's seat, monitor the running status of the vehicle and the surrounding environment, and be ready to take it over at any time. When the vehicle gives a takeover request or is in a state not suitable for autonomous driving, the user-in-charge must immediately take it over.[20] Furthermore, Article 53 of the Shenzhen Regulations confirms that the driver of a level 3 or level 4 automated vehicle should assume the liability for compensation where he is held liable for a traffic accident.

Regarding the breach of duty issue, the driver's liability for traffic accidents caused by traditional cars is negligence-based. Specifically, the driver will not be held liable for compensation of the traffic accident victims unless he is guilty of negligence, that is, failing to meet the standard of a reasonable driver in the circumstances. The current law requires the victim to have a burden of proof to establish both breach and causation elements, although, in practice, the evidence produced by the traffic police officers after their investigation is critical for determining these two legal issues.

In principle, the same legal principles regarding breach of duty and causation may be applied by analogy to the user-in-charge of an automated vehicle. Similarly, the user-in-charge, as required as a competent and careful driver, must be qualified to drive with a driving license, keep physically and mentally fit for driving, maintain the vehicle in a roadworthy condition, timely respond to a transition demand, properly monitor

[19] The *Civil Code* (民法典) was made by the National People's Congress on 28 May 2020, effective on 1 January 2021, accessed 19 Feb 2024.

[20] See Art.35(1) of Shenzhen Regulations.

the driving environment and respond by using the vehicle controls to steer, brake, slow down, accelerate, turn on lights or indicate appropriately, and comply with the road safety regulations.[21] Nevertheless, the user-in-charge will be free of negligence if he fails to meet the above requirements because he suffers a sudden medical condition neither caused nor reasonably foreseeable by him. The extant legal principles on causation are equally applicable to the case of automated vehicle accidents.

However, the issue of whether the victim of automated car accidents should bear the burden of proof on the breach of duty and causation issues is open for discussion. Compared to traffic accidents caused by those in traditional cars, the issues of breach and causation in automated vehicle accidents are more complicated because low-level automated vehicles may be operated either through an automatic control system or under the user-in-charge's manual control at the time of the accident. In other words, the victim has a heavy burden to ascertain the exact technical reasons for the occurrence of a traffic accident and to establish that the accident was caused by the negligent operation of the user-in-charge when the car was being driven in a manual control mode (Kouroutakis, 2020, p. 1115).[22]

According to Articles 70, 72, and 73 of the *Road Traffic Safety Law*, the parties to a traffic accident must report the case to the traffic administrative department of the local public security authority, which should immediately send traffic police officers to the scene of the accident, inspect it and collect evidence, and then produce a written conclusion of the inspection or an inspection report of the traffic accident in time. The written conclusion should include basic facts, the causes of the traffic accident, and the liabilities of the parties concerned. Similarly, Article 52(1) of the Shenzhen Regulations provides that the user-in-charge of the automated vehicle must immediately park the car and protect the scene; if the accident causes personal injury or death to a victim, the driver must immediately rescue the victim and promptly report this to traffic police officers. However, suppose traffic police officers cannot conclude due to a lack of evidence or other reasons. In that case, the victim must bear the additional expenses of seeking evidence on complicated technical issues relating to

[21] See Law Commission of England and Wales and Scottish Law Commission, 'Automated Vehicles: Joint Report' (2022) https://www.lawcom.gov.uk/document/automated-vehicles-final-report/ accessed 19 Feb 2024.

[22] Germany's Road Transportation Act amended in 2017 requires a "black box" to be installed in automated vehicles to get more information and clarify the conditions of traffic accidents.

the user-in-charge's potential negligence regarding driving an automated vehicle and a causal link between the negligence and the damage in the traffic accident. Otherwise, he must bear the losses himself because the current law imposes the burden of proof on him.

Because of an obvious information asymmetry between the victim and the user-in-charge in terms of the functions and performance capacities of the automated vehicles concerned, the author proposes relaxing the victim's burden of proof regarding breach of duty and causation—a feasible legal solution is reversing the burden of proof on the negligence and causation issues to the user-in-charge. Such a suggestion will bring the Chinese legal approach closer to German law in terms of the determination of the driver's liability for automated vehicle accidents—the driver is jointly liable with the keeper (who bears strict liability for an accident) unless the driver can prove he neither cause the accident intentionally nor negligently (Patti, 2019, p. 134).

The author's suggestion derives from the existing legal principle of reversing the burden of proof under Chinese tort law, which reverses the burden of proof on the causation issue in environmental pollution and ecological damage cases. As Article 1230 of the *Civil Code* provides, where any dispute arises from environmental pollution or ecological damage, the actor should bear the burden to prove that he should not be liable or that his liability could be mitigated as provided by law and that there is no causation between his act and the damage. The law reverses the burden of proof because the plaintiffs are substantially weaker than the defendants in obtaining evidence in environmental pollution and ecological damage cases. By the same token, when the victims of a traffic accident caused by an automated vehicle with a user-in-charge are pedestrians, drivers, or passengers of non-motor vehicles or traditional cars, they are also in a weaker position than the user-in-charge of automated vehicles in terms of obtaining evidence regarding the breach and causation issues. Therefore, the author suggests that the law adopt the same approach and require the user-in-charge to bear the burden of proof for breach and causation in traffic accidents caused by automated vehicles. However, in the author's view, when a traffic accident occurs between two automated vehicles and the victim of the driver of one of the crashed vehicles, the principle of reserving the burden of proof should not apply, and it remains for the plaintiff to establish breach and causation.

In addition, a technical solution to mitigate information asymmetry is installing a "black box" in automated vehicles. This solution has been widely used in aircraft for facilitating the investigation of aviation accidents. For example, Section 63a of the 2017 *Amended Road Traffic Act of Germany* provides that an automated vehicle should record the geographical position and time data when the control of the vehicle changes from the driver to the automated system or vice versa and the data should also be stored when the system prompts the driver to retake control of the vehicle or when the system experiences a technical fault. The author suggests that Chinese law should take the German new requirement as a reference and adopt such a technical solution to tackle the evidentiary difficulty regarding breach of duty and causation in automated vehicle accidents.

Restrictions on Civil Liabilities of a User-in-Charge for Car Accidents

Article 76 of the *Road Traffic Safety Law* provides the legal principles of comparative negligence in traditional traffic accidents by distinguishing the type of victim. Specifically, where a traffic accident occurs between motor vehicles, the party at fault should bear the liabilities. If both parties are at fault, they should each take their appropriate share of the liabilities. Where a traffic accident occurs between a motor vehicle and a non-motor vehicle driver or a pedestrian, if there is any evidence to prove that the non-motor vehicle driver or the pedestrian is at fault, the motor vehicle driver's liabilities may be appropriately mitigated considering the degree of fault. Still, if the motor vehicle is not at fault, the motor vehicle driver shall bear no more than 10 per cent of the liability for compensation.

Although the lawmakers of the *Road Traffic Safety Law* did not consider traffic accidents caused by automated vehicles, Article 76 of the law may be applied by analogy as follows. When a traffic accident occurs between motor vehicles (regardless of whether the victim is inside a traditional or automated vehicle), the user-in-charge and the victim at fault must bear liability for compensation according to the degree and contribution of their fault to the accident. This is confirmed by Article 1173 of the *Civil Code*, which generally stipulates that the tortfeasor's liability may be reduced where the tort victim is at fault regarding the occurrence or aggravation of the harm. By contrast, where a traffic accident occurs between a low-level automated vehicle and a pedestrian and a non-motor vehicle, the principle of contributory negligence set out in Article 76 of the *Road Traffic Safety Law* should apply: (1) the user-in-charge may request a reduction in the amount of compensation when the victim is at fault either through contributing to the accident or the damage; (2) where the user-in-charge is not at fault, he is still required to share no more than 10 per cent of the amount of damages suffered by the victim.

Highly Automated Vehicles without a User-in-Charge

In an automated car with high-level driving automation (such as a level 5 car), there is no driver or user-in-charge who operates the driving controls. They are thus predicted to shift from a traditional product-oriented market to a service-oriented passenger vehicle market. Such automated cars are specifically regulated by the 2021 *Autonomous Driving Act of Germany*, which defines "vehicles with autonomous driving capabilities" as motor vehicles that "can perform driving tasks independently without a person driving" in specified operating areas on public roads.[23] Since there is no driver in a highly automated vehicle, the civil liabilities for an automated car accident in a driverless situation essentially differ from those in a traditional car accident.

Three major issues arise when such automated vehicles cause traffic accidents and harm various victims, whether they are road users outside the car or passengers inside the vehicle. (1) What is the nature of the tortious liability in this case—is it a liability for a traffic accident or product liability? (2) Who are the potentially liable parties for compensation? (3) To what extent do the extant legal principles regarding product liability need to be adapted to the uniqueness of automated vehicles? This part is to discuss these issues in detail.

A Shift from Accident Liability to Product Liability

Regarding the first issue, the essential difference between traffic accident liability and product liability is two-fold. First, liability for a traffic accident occurs when a motor vehicle driver fails to meet the standard of a

[23] 'Germany: Road Traffic Act Amendment Allows Driverless Vehicles on Public Roads', *Library of Congress*, 9 Aug 2021, https://www.loc.gov/item/global-legal-monitor/2021-08-09/germany-road-traffic-act-amendment-allows-driverless-vehicles-on-public-roads/ accessed 19 Feb 2024.

reasonable driver and thus causes damage to other road users. By contrast, product liability occurs when the manufacturer or the seller makes or sells a defective product and thus causes the consumer to suffer personal injury and/or property damage. Second, under Chinese law, liability for a traffic accident is negligence-based, while product liability is strict liability, and the fact of whether the actor is at fault or not is irrelevant. However, the law provides some circumstances of exemption from product liability.

Based on the above comparison between traffic accident liability and product liability, in the author's opinion, when a driverless automated vehicle is operated through an automatic control system, and a traffic accident occurs due to a defect associated with the hardware structures or software systems of the vehicle, product liability is a proper cause of action for car accident victims claiming compensation.[24]

Re-defining "Manufacturers" and Automated Vehicles as a "Product"

In terms of the second issue (i.e., who are the potentially liable parties for compensation?), where the victims of car accidents claim compensation for product liability, the potentially responsible parties include the manufacturer and the seller of the automated vehicle under the current Chinese law. The general legal rules regarding product liability are in Chapter Four, titled "Product Liability" of the *Civil Code*. Specifically, Article 1202 of the *Civil Code* provides that where a defective product causes damage to another person, the manufacturer should bear tort liability. Articles 1203 and 1204 of the *Civil Code* further provide that the injured party is entitled to claim compensation either against the manufacturer or the seller of the product as both are jointly liable for the product defect regardless of which party caused the defect and whether or not the defect resulted from the fault of a third party (such as a transporter or a warehouse). Moreover, according to Article 1206 of the *Civil Code*, where a product defect is discovered after the product is put into circulation, the manufacturer or the seller should take remedial measures (such as stopping sales, providing warnings, or recalling the product) in a timely manner. Suppose they fail to take remedial measures promptly or take ineffective measures to aggravate the damage. In that case, they should be liable for

[24] The same analysis applies to cases where a traffic accident occurs due to a defect in an automated vehicle with a user-in-charge.

the aggravated part of the damage. Article 1207 of the *Civil Code* also requires the manufacturer and the seller to pay punitive damages to the injured party where they manufacture or sell a product knowing that the product is defective or they fail to take remedial measures as required by law as a result of which the injured party suffers death or serious personal injury.

In principle, the above legal principles of product liability apply to automated vehicles. However, the unique features of automated vehicles require the general principle to be adjusted accordingly.

Unlike traditional cars, automated vehicles combine the hardware structures, the software systems operating in the entire industry-grade toolchains, and the network-based data-driven information system. The hardware and software of automated vehicles are not necessarily released at the same time. Hence, the concept of "manufacturers" of automated vehicles should be reinterpreted: it should include not only the manufacturer of the hardware structures but also the provider of the complicated software systems embedded in automated vehicles. In other words, the latter's post-circulation upgrade of the embedded software systems for wheels with the over-the-air upgrade capacity should still be seen as part of the manufacturing process of automated vehicles. Moreover, connected automated vehicles must run on data that revolve around a sophisticated network-based system that collects traffic-related information from inside and outside the cars. Therefore, the controller and processor of traffic-related information systems should also be regarded as a manufacturer of automated vehicles.

Moreover, the coverage of the concept of "defects" in automated vehicles needs to be expanded—it should not be limited to any defect of the physical hardware structures (such as a defect in the sensors installed in automated vehicles). Instead, it should include defects in software systems embedded in self-driving vehicles (such as a design defect in the software operating system which leads to a failure to respond to a particular traffic risk or crashed software effectively) (Kim, 2018, p. 303) and a defect arising from the network-based data-driven information systems (such as a defect relating to data accuracy and data sharing or a defect resulting in a cybersecurity vulnerability). To determine whether an automated vehicle is free of defects, the hardware structures, the software systems, and the information systems must be up to the relevant quality standards set by the law or the industry.

Articles 1206 and 1207 of the *Civil Code* state that whenever manufacturers discover any defect in automated vehicles after the cars are sold in the market, they must take adequate remedial measures such as stopping sales, providing warnings, or recalling the cars in a timely manner.[25]Otherwise, they are required to compensate for the aggravated part of the damage. They may be ordered to pay punitive damages if death or serious personal injury is caused to the victims of traffic accidents.

Adaption of the Law of Product Liabilities to Automated Vehicles

Concerning the third issue (i.e., to what extent do the extant legal principles regarding product liability need to be adapted to the uniqueness of automated vehicles?), the following four aspects are worth discussing: the joint product liability of multiple manufacturers; the burden of proof for establishing product liability; the applicability of the statutory defences including the state-of-the-art defence; and the liability of the owners or controllers of automated vehicles.

1. The Joint Product Liability of Multiple Manufacturers

According to Article 1202 of the *Civil Code*, all the manufacturers of automated vehicles should jointly bear strict liability for a product defect that causes a traffic accident, regardless of which manufacturer caused the defect and whether or not the defect resulted from the fault of a third party (such as hackers), which is similar to the "common enterprise liability" suggested by Professor Vladeck (Vladeck, 2014, p. 149). To successfully claim compensation, the victim only needs to prove that he has suffered damages in a traffic accident caused by a defective automated vehicle, normally by using the written conclusion of traffic police officers after their inspection and investigation of the traffic accident. In other words, the victim of the traffic accident can claim compensation from any of the manufacturers, including the manufacturer of the hardware structures, the software system provider, and the controller and processor of the information systems. Among the multiple manufacturers who bear joint and several liabilities, the allocation of the liability should be decided according to the degree of contribution by the parties responsible for the defect. Because of the complicated hardware and software combination

[25] Art.24 of Shenzhen Regulations.

and the complex algorithmic technologies of automated vehicles, joint and several liability mechanisms help the victim avoid the practical difficulties of ascertaining the responsible party for the defect that caused the traffic accident. Moreover, Article 1203 of the *Civil Code* requires the seller of an automated vehicle to bear joint and several liabilities with the manufacturers, which enables the victim to quickly identify the defendant and target more liable parties in his claim for compensation.

2. The Burden of Proof for Establishing Product Liability

Although the current Chinese law requires the victim of a defective product to bear the burden of proof in relation to two issues—the product's defect and the causation between the defect and the damage—it is extremely difficult and costly for laypeople victims to prove a defect of an automated vehicle and its causal link to the consequent traffic accident given the complicated technologies of automated vehicles and the algorithmic decision systems used therein. As noted in Part Three above, Chinese law reverses the burden of proof on the causation issue to the defendant in environmental pollution and ecological damage cases to facilitate the victims to obtain legal remedies.[26] The victims of defective automated vehicles are in a similar situation and are unlikely to receive compensation where the law imposes on them the heavy burden of proof concerning the defect and causation issues. By contrast, the defendant manufacturers and sellers of automated vehicles not only have significant advantages in obtaining complicated technical evidence but also have deeper pockets and are in a better position to distribute the risks and losses in society. Therefore, the author suggests that Chinese law should consider shifting the burden of proof regarding the defect and causation to the manufacturers and the sellers of automated vehicles. This would encourage the automated vehicle industry to enhance car safety and cultivate consumer confidence by allocating the legal risks and losses associated with the complexity and uncertainty of automated vehicle technologies to the industry rather than vulnerable victims.

3. The Applicability of the Statutory Defences

[26] See the discussion in the Section above titled "Establishment of Civil Liabilities of a User-in-Charge for Car Accidents".

Article 41 of the *Product Quality Law*[27] sets out three circumstances for exempting the manufacturer from product liability: (1) the product has not been put into circulation; (2) the defect concerned was non-existent when the product was put into circulation; and (3) the defect concerned could not be found at the time of circulation due to scientific and technological reasons (i.e., the "state-of-the-art defence"). The manufacturer should have the burden of proof on the exemptions. However, the author argues that the above state-of-the-art defence should not apply to the case of defective automated vehicles.

First, unlike traditional vehicles, the safety of automated vehicles heavily relies on their performance at the stage of road testing, during which massive training data are collected, processed, and used for improving steering, braking and various functions independently played by autonomation systems. The risk of automated vehicle accidents before the circulation cannot be excluded from the coverage of product liability. Therefore, the current local laws on automated vehicles specifically regulate the stage of road testing and practical demonstration of automated vehicles before they are put into circulation and, in doing so, nullify the first circumstance of exemption. For example, Article 49(2) of the Shanghai Measures provides that the victim is entitled to claim compensation against the manufacturers or the sellers where intelligent connected vehicles cause a traffic accident or other damage due to their product defect. Article 54 of the Shenzhen Regulations also confirms the product liability of the manufacturers and sellers of automated vehicles at this stage.

Second, the second and third circumstances exempt the manufacturer from product liability for any defect non-existent or unknown due to the state-of-the-art at the time of circulation. However, the software systems embedded in automated vehicles need to be regularly upgraded even after vehicles are sold to consumers. Moreover, software-oriented automated vehicles must operate on network-based data-driven structures. In other words, the manufacturers (primarily the provider of the software systems and the controller and processor of the traffic-related information systems) continue to "manufacture" the product of automated vehicles after they are sold in the market. Different from traditional cars, the

[27] The *Product Quality Law* (产品质量法) was made by the Standing Committee of the National People's Congress on 22 February 1993, effective 1 September 1993, amended in 2000, 2009, and 2018, https://gkml.samr.gov.cn/nsjg/fgs/201906/t20190625_302770.html accessed 19 Feb 2024.

manufacturers of automated vehicles are required to detect and then correct the defects which occur after the cars are put into circulation, although they are unknown to them at the time of circulation. In addition, Article 1206 of the *Civil Code* requires the manufacturer or seller to take remedial measures promptly when a defect in a product is discovered after the product is put into circulation, which means that their product liability has extended over the time of circulation. Therefore, the author suggests that the manufacturers of automated vehicles should be responsible for non-existent or unknown defects at the time of circulation,[28]which in turn would help to enhance consumer confidence in using automated vehicles.

4. The Liability of the Owners or Controllers

Another controversial legal issue is whether the owners or controllers of defective automated vehicles should also be responsible for traffic accidents caused by defects in automated vehicles. If affirmative, do they need to bear joint or residual liability to the victim of defective automated vehicles?

To address the question of whether the owners or controllers of automated vehicles should be responsible for product liability, the author suggests that it be considered and answered by distinguishing the nature and type of the owners or controllers. If an owner or controller of an automated vehicle is an individual consumer who purchased an automated vehicle for personal use, he should not be liable for a traffic accident caused by a defective automated vehicle unless he has misused the car and contributed to the occurrence of the traffic accident. By contrast, if the owner or controller of an automated vehicle is a business enterprise that purchases and uses automated vehicles for a commercial purpose (such as providing transportation services to customers), it has a central operation center to oversee all driverless automated vehicles in its business and is capable to bear liability for compensating the victim whenever a defect in its automated vehicle causes a traffic accident.

However, Article 53(2) of the Shenzhen Regulations requires that where a traffic accident is caused by a reason associated with a level 5 automated vehicle, the owner or controller of the automated vehicle should first bear the liability for compensation. They are entitled to exercise the

[28] However, the damages occurring between the time of circulation and the time of discovering a defect should be excluded from the scope of product liability.

right of recourse against the manufacturer or the seller if any defect caused the accident in the automated vehicle according to Article 54 of the Shenzhen Regulations. By the same token, the driver of a level 3 and level 4 automated vehicle is also required to vicariously bear the liability for compensating the victim for the manufacturer or the seller of the car in the case of product liability.[29]

As analyzed above, the author suggests that these provisions should be narrowly interpreted as applying only to a commercial driver, owner, or controller of an automated vehicle because it is unreasonable to impose vicarious product liability on a consumer-deriver, owner, or controller of an automated vehicle for three reasons. First, the law normally imposes vicarious liability on a party because it has deeper pockets with which to compensate victims, it benefits from an activity as a result of which a tort has been committed, and it creates the risk of the tort. To some degree, it has control over the tortfeasor. However, these elements do not apply to the consumer-driver, owner, or controller of an automated vehicle in product liability cases. Second, an automated vehicle's consumer-driver, owner, or controller lacks the means and resources to fix its defects. Third, imposing vicarious product liability on the consumer-driver, owner, or controller of an automated vehicle will stifle consumer demand and choice of automated vehicles because of the excessive legal risks and burdens they face (Hoong, 2019, p. 32). This will further harm the common interests of the automated vehicle industry. Currently, Articles 53 and 54 of the Shenzhen Regulations may not immediately cause problems because, at the stage of road testing and practical demonstrations, the drivers, owners, or controllers of automated vehicles are those companies that are testing and developing automated vehicles and who will become the manufacturers when wide use of automated vehicles is achieved. However, the law should not require the consumer-driver, owner, or controller of automated vehicles to bear joint product liability for the manufacturer and the sellers.

Moreover, in the case of the commercial owners or controllers of automated vehicles, the author suggests that they should bear "residual" rather than "joint" liability for defective vehicles with the manufacturers and sellers because product liability is primarily a type of tortious liability of the automated vehicles industry which has the capacity and is in the best position to enhance the vehicles' safety and functions. The commercial owners

[29] Art.53(1) and Art.54 of the Shenzhen Regulations.

or controllers of automated vehicles are users only, and there is a lack of justification for holding them jointly liable for product liability for defective automated vehicles.

On the other hand, the commercial owners or controllers of automated vehicles may be seen as "the operator or manager of commercial or public premises" set out in Article 1198 of the *Civil Code*. As Article 1198(2) provides, the operator or manager of commercial or public premises, if failing to fulfil the duty of safety protection, should assume the residual liability when the third-party direct tortfeasor causes harm to the victim on commercial or public premises and is unable to compensate the victim fully. At the same time, the operator or manager has the right of recourse against the third-party direct tortfeasor. When highly automated vehicles providing transportation services to passengers are arguably "commercial or public premises", it is appropriate for commercial owners or controllers of such automated vehicles that fail to discharge the duty of safety protection to assume residual liability for traffic accidents caused by defects in automated vehicles when the direct tortfeasors (i.e., the manufacturers and sellers) cannot fully compensate the victim.

Liability Insurance Models for Automated Vehicles

Regardless of being traditional or automated, each vehicle must carry insurance. Insurance is indispensable and essential for spreading the risks of legal liabilities and distributing losses, thus incentivizing the automated vehicle industry and, at the same time, enhancing consumer confidence in using automated vehicles.

However, the liability insurance models for automated vehicles are more complicated than that applied to traditional ones, where the law normally requires the drivers to purchase third-party insurance. As analyzed above, civil liabilities for traffic accidents caused by automated vehicles can be traffic accident liability when a user-in-charge negligently causes the accident or product liability when the manufacturers and sellers contribute to a defect in an automated vehicle. Under Chinese law, three types of insurance and one alternative are associated with automated vehicles.

The first type is compulsory motor vehicle traffic accident liability insurance. For example, Article 28 of the Shenzhen Regulations and Article 40(1) of the Shanghai Measures require the controller or owner of an automated vehicle to purchase compulsory motor vehicle traffic accident liability insurance. When an insured automated vehicle causes a traffic accident, regardless of whether the driver is negligent or not, the insurance company should indemnify the losses resulting from personal injury or property damage of the victims other than the passengers and the insured within the limit of indemnification unless the victims intentionally cause the accident. Although the *Regulation on Compulsory Motor Vehicle Traffic Accident Liability Insurance*[30] applies to traditional motor vehicles, its application scope may extend to automated vehicles for the same purpose of ensuring the indemnification of the victims in road motor vehicle traffic accidents and promoting road traffic safety.

The second type of insurance is commercial third-party motor vehicle liability insurance for the driver's liability for the negligent driving of level-3 automated vehicles. For example, Article 28 of the Shenzhen Regulations also requires the controller or owner of an automated vehicle to purchase third-party motor vehicle liability insurance. Moreover, because either compulsory motor vehicle traffic accident liability insurance or commercial third-party motor vehicle liability insurance does not indemnify the passengers inside the vehicles, Article 28 of the Shenzhen Regulations further requires the controller or owner of an automated vehicle with passenger-carrying functions to purchase liability insurance for passengers. Similarly, Article 40(1) of the Shanghai Measures also provides that the controller of an automated vehicle should buy commercial insurance or a letter of guarantee for a certain amount of money.

While the first two types insure the driver-based liability, the third type of insurance associated with automated vehicles is product-based liability insurance. For traffic accidents caused by any defect in an automated vehicle, the potentially liable parties, including the manufacturers and the sellers of automated vehicles, need to purchase product liability insurance. Indeed, a collective product liability insurance arrangement similar to "manufacturer enterprise responsibility", as suggested by Professor Abraham and Professor Rabin (Abraham & Rabin, 2019, p. 147), suits well the Chinese legal context where multiple manufacturers and sellers of automated vehicles are required to jointly bear liability for compensating injured parties in a traffic accident following an autonomously performed

[30]The *Regulation on Compulsory Motor Vehicle Traffic Accident Liability Insurance* (机动车交通事故责任强制保险条例) was made by the State Council on 21 March 2006, effective 1 July 2006, amended in 2012, 2016, and 2019, http://www.gov.cn/gongbao/content/2019/content_5468886.htm accessed 19 Feb 2024.

driving task. Although Article 10(1) of the Shenzhen Regulations only encourages the insurance industry to develop a full set of insurance policies that cover the design, production, use, management, algorithmic service, and other product risks of intelligent connected vehicles, Chinese law should consider a mandate for automated vehicle manufactures and sellers' product liability insurance because of a relatively higher risk of product liability than a driver-based liability in the case of an automated vehicle crash (Patti, 2019, p. 140). Moreover, Article 40(2) of the Shanghai Measures also encourages the insurance industry to innovate by developing new insurance policies that fit the features of connected automated vehicles, such as an insurance policy against the risk of hackers.

Finally, as an alternative to insurance, the State may establish a social assistance fund or the automated vehicle industry may establish a social risk fund. Article 24 of the *Regulation on Compulsory Motor Vehicle Traffic Accident Liability Insurance* establishes a social assistance fund for traffic accidents caused by traditional motor vehicles to reimburse the victims where the compulsory motor vehicle traffic accident liability insurance is unavailable or cannot cover the losses (Pearl, 2019, p. 1870).[31]Although it does not cover automated vehicles, the author suggests a similar social assistance fund may be considered when automated vehicles become more widely used. At the current stage of road testing and practical demonstrations of automated vehicles, establishing a social risk fund established by the automated vehicle industry is a better choice. For example, Article 40(3) of the Shanghai Measures provides that the law encourages the relevant association and companies manufacturing intelligent connected vehicles to jointly establish a social risk fund, which can reimburse the victims of traffic accidents caused by intelligent connected cars when they cannot get prompt compensation due to unascertainable liabilities or other reasons.

Conclusion

Low-level automated vehicles have been gradually put into the market. Highly automated vehicles, especially level 4 and level 5 ones, are the future of transport. However, as in other countries, software-oriented connected automated vehicles have raised legal challenges in China. At the current stage of road testing and practical demonstrations of automated

[31] This reassembles the so-called "public victim compensation funds".

vehicles, it is urgent and essential to explore a suitable legal framework of civil liability for automated vehicle accidents. The basic strategies that China can adopt are applying the current road traffic safety law and product liability law, with appropriate adjustments and adaptions, to deal with these challenges of automated cars, and at the same time, encouraging local governments to explore and experiment with innovative legal solutions to overcome the challenges so that the Central Government can compare and assess their effectiveness and use them as an essential reference for making a national law that comprehensively regulates automated cars. This chapter distinguishes automated vehicles with a user-in-charge and highly automated vehicles without a user-in-charge for legal analysis. It presents a basic legal framework that includes legal principles regarding traffic accident liability, product liability, and potential diverse insurance models for automated vehicle accidents. With a broader range of commercial applications of automated vehicles and a better understanding of the associated legal risks, these proposed legal principles will be tested in practice and revisited for improvement and specification shortly.

References

Abraham, K. S., & Rabin, R. L. (2019). Automated vehicles and manufacturer responsibility for accidents: A new legal regime for a new era. *Virginia Law Review, 105*(1), 127–171.

Carp, J. A. (2018). Autonomous vehicles: Problems and principles for future regulation. *University of Pennsylvania Journal of Law & Public Affairs, 4*(1), 81–148.

Feng, J. 冯纪朋. (2021). The status quo of China's development of automated vehicles and the analysis on the reform of comprehensive transportation system (我国自动驾驶发展现状及对综合交通体系的变革分析). Communication & Shipping 3, 62.

Hoong, R. (2019). Autonomous vehicles and torts: A behavioral economic analysis. *Pennsylvania Undergraduate Law Journal, 12*(1), 32–49.

Kim, S. (2018). Crashed software: Assessing product liability for software defects in automated vehicles. *Duke Law & Technology Review, 16*(1), 300–317.

Kouroutakis, A. E. (2020). Autonomous vehicles: Regulatory challenges and the response from Germany and UK. *Mitchell Hamline Law Review, 46*(5), 1103–1121.

Patti, F. P. (2019). The European road to autonomous vehicles. *Fordham International Law Journal, 43*(1), 125–162.

Pearl, T. H. (2019). Compensation at the crossroads: Autonomous vehicles & alternative victim compensation schemes. *William & Mary Law Review, 60*(5), 1827–1891.

Vladeck, D. C. (2014). Machines without principals: Liability rules and artificial intelligence. *Washington Law Review, 89*, 117–150.

CHAPTER 8

Data Governance Rules in the Context of Connected Vehicles

Weixing Shen and Yun Liu

Introduction

The development of autonomous driving technology is heavily reliant on extensive data from the road environment. The operation and maintenance of autonomous vehicles are contingent upon the recording and transmission of data from vehicle equipment. As autonomous vehicles are increasingly deployed and in constant motion, the extent and scale of data collected by these vehicles are expanding continuously. This has resulted in the governance of vehicle data becoming a pressing concern among various jurisdictions.

For instance, in February 2024, the U.S. Department of Commerce issued a notice regarding specialized rules for connected vehicle data to

W. Shen
School of Law, Tsinghua University, Beijing, China
e-mail: wxshen@tsinghua.edu.cn

Y. Liu (✉)
Think Tank Center, Tsinghua University, Beijing, China
e-mail: liuyun89@tsinghua.edu.cn

C. Ding, R. Fan (eds.), *Chinese Ethical and Legal Perspectives on Automated Vehicles*,
https://doi.org/10.1007/978-981-96-1277-2_8

ensure the security of the information and communication technology and services supply chain for connected vehicles.[1]Similarly, the European Data Protection Board (EDPB) released guidelines on processing personal data in connected cars and mobile applications. In China, the "Several Provisions on the Management of Automobile Data Security (for Trial Implementation)" was released in 2021, requiring companies involved in vehicle data to submit an annual report on data security. Elon Musk, the founder of Tesla, visited Beijing to address issues related to the security and compliance of vehicle data.[2]

Furthermore, there are distinct regulations focused on vehicle data, such as the "Beijing Autonomous Driving Data Categorization and Classification Demonstration 2.0," the "Measures for the Management of Data Security in the Beijing Intelligent Connected Vehicle Policy Pioneer Zone (for Trial Implementation)," and the "Regulations on the Administration of Automobile Data Processing in Zhejiang Province." In China, entities subject to overseeing the security of vehicle data encompass not only automobile manufacturers but also ride-hailing platforms, map sensor providers, and other entities whose operational frameworks are significantly reliant on vehicle data.

This paper focus on the key legal issues involved in the governance of data related to autonomous driving.

Data Processing Practices and Emerging Risks in Vehicle Connectivity

Vehicles equipped with autonomous features must possess the capability for continuous data recording and sharing. In the regulations governing autonomous vehicle testing in China, the European Union, and the United States, manufacturers are required to install data recording and storage devices in their vehicles. The data generated during road tests and autonomous vehicle operations must be securely stored, and information pertaining to accidents and other incidents must be promptly reported to

[1] "Securing the Information and Communications Technology and Services Supply Chain: Connected Vehicles." Bureau of Industry and Security, U.S. Department of Commerce, [Docket No. 240227-0060], [RIN 0694-AJ56], Federal Register, 15 CFR Part 7, available at https://www.federalregister.gov/d/2024-04382

[2] Kang, Ansuya. "Musk Visits Beijing as Tesla's China-Made Cars Pass Security Rules." CNBC, 29 Apr. 2024, available at www.cnbc.com/2024/04/29/musk-visits-beijing-as-teslas-china-made-cars-pass-security-rules.html

relevant authorities. As autonomous vehicles gather and analyze larger volumes of data, the issue of data security has become a focal point of concern. Globally, a range of challenges have surfaced, including instances of vehicle security breaches by ethical hackers during operations, unauthorized disclosure of personal information stored in vehicles, and imposing overly strict regulatory measures on the cross-border transmission of vehicle data, resulting in restrictions on the trade activities. In particular, we can examine the security risks and regulatory requirements related to vehicle data governance across three distinct types of data processing activities.

Data of in-Vehicle Occupants

In autonomous vehicles, in-vehicle sensors designed to record driver and passenger behavior information primarily include the Driver Monitoring System (DMS) and the Occupant Monitoring System (OMS). Collectively known as cabin monitoring systems, these technologies are integral for enhancing driver and passenger safety, optimizing autonomous driving system performance, and meeting regulatory requirements.

The DMS monitors real-time information such as facial features, eye movements, and head posture to accurately assess the driver's fatigue and attention levels. This capability is crucial for preventing dangerous behaviors like drowsy and distracted driving. When the DMS detects fatigue or distraction, it promptly issues alerts and reminders to encourage safe driving, thereby significantly reducing accident risks associated with poor driver conditions.[3]

The OMS oversees the presence and safety status of passengers within the vehicle, enhancing the functionality of safety systems like airbags and seatbelts to provide optimal protection in emergency situations. Moreover, the OMS is capable of identifying critical scenarios, such as condition of child safety seats or potential theft incidents, and promptly sending alerts to protect the lives and belongings of the passengers.

In situations where the autonomous driving system is unable to completely manage the vehicle, such as in complex traffic conditions or emergencies, the DMS and OMS enable quick intervention by the driver or

[3] Performance Requirements and Test Methods for Driver Attention Monitoring System(GB/T 41797-2022), National Automotive Standardization Technical Committee (SAC/TC 114), date of issue:2022-10-14, date of implementation:2023-05-01.

passengers to avert accidents. Thanks to advancements in machine vision and artificial intelligence, the capabilities of DMS and OMS are constantly evolving, enabling more thorough monitoring and interaction with onboard systems. This, in turn, enhances personalized settings and comfort levels for a superior onboard experience.

In recent years, new legislation has been introduced to enhance vehicle safety performance. To mitigate traffic accident risks and improve the safety of drivers and passengers, numerous jurisdictions have enacted regulations and standards requiring the installation of DMS and OMS in vehicles. For instance, China imposed mandatory requirements for DMS installation in commercial vehicles such as hazardous materials transporters and passenger carriers in 2018.[4] Ride-hailing platforms in China also provide voice prompts indicating that in-car environments are being recorded upon passenger entry to prevent disputes. From July 2024, the European Union require all new cars entering the European market to be equipped with driver fatigue monitoring systems. As technology and relevant legal frameworks evolve, DMS and OMS are already becomed standard features in future vehicles.

However, the continuous collection and storage of information or data by vehicle remote monitoring systems pose significant privacy risks. These systems can systematically analyze driving habits, travel trajectories, destinations visited, in-car conversations, and facial features, exposing drivers to vulnerabilities in the digital realm. Unreasonable use of such information or data can infringe on drivers' privacy and personal information rights. Several incidents of autonomous vehicle cameras being remotely controlled and sensitive in-car photos being leaked have been reported by Chinese media in recent years. In October 2021, the National Information Security Standardization Technical Committee issued the "Guidelines for the Safe Handling of Vehicle Collected Data," recognizing the vehicle cabin as a private space for drivers and passengers, consistent with privacy provisions in the Civil Code. To protect vehicle data and user privacy, numerous car manufacturers, including Dongfeng Nissan, BYD, XPeng Motors, Kia, HiPhi, Tesla, Great Wall Motors, Li Auto, NIO, and WM Motor, have repeatedly shut down, disabled, removed, adjusted, or upgraded their remote monitoring functions throughout 2022.

[4] Ministry of Transport of the People's Republic of China, "Notice of the General Office of the Ministry of Transport on the Promotion and Application of Intelligent Video Monitoring and Alarm Technology." August 22, 2018.

Data of Vehicle as the Machine

Autonomous vehicles also capture a vast array of fundamental vehicle information and associated operational data, which can prove invaluable in various scenarios. Firstly, essential vehicle data, such as vehicle identification details (like frame number or license plate information), vehicle type, manufacturer, etc., plays a crucial role in tracing vehicle identity, particularly in contexts like used car transactions. Secondly, data concerning changes in control modes, which document transitions between different driving modes (e.g., manual driving, assisted autonomous driving, fully autonomous driving), is essential for evaluating the timing and efficacy of interventions by autonomous systems. Thirdly, records of received remote control commands are maintained; in cases where an autonomous vehicle supports remote control, all commands received remotely are recorded to ensure operational traceability and safety. Lastly, information on vehicle operational status, encompassing factors like location, speed, acceleration, travel direction, as well as real-time lighting and signaling status, is vital for real-time vehicle monitoring and assessment of driving behaviors.

China has comprehensive regulations governing the real-time recording and uploading of operational data pertaining to new energy vehicles. For instance, the "Technical Specification for Remote Service and Management System of Electric Vehicles"(GB/T 32960-2016) issued by the Standardization Administration of China details technical requirements for electric vehicle remote service and management systems, including data collection, storage, and transmission. Specifically, this standard mandates that new energy vehicles transmit real-time data to a backend system at predetermined intervals (e.g., every 30 seconds) for safety purposes. This data includes essential information such as vehicle location, speed, battery status, and motor status. Some local regulations also stipulate this. For example, the "Jiangsu Province Road Traffic Safety Regulations" and the "Shenzhen Special Economic Zone Regulations on the Administration of Intelligent and Connected Vehicles" both require conditionally autonomous, highly autonomous, and fully autonomous vehicles to record and store at least 90 seconds of data prior to an incident or malfunction, including location, operational status, driving mode, interior and exterior surveillance videos, etc., maintaining data continuity and integrity, with a storage period of no less than 30 days.

The United Nations Economic Commission for Europe (UNECE) has issued technical regulations requiring vehicles to be equipped with a "black box" data storage system (DSSAD) to record various data when the Automated Lane Keeping Systems (ALKS) is activated, facilitating data retrieval in case of an accident.[5] The Ministry of Land, Infrastructure, Transport and Tourism (MLIT) in Japan mandates that autonomous vehicles must be equipped with systems capable of recording and storing data related to vehicle control, sensor input, and operational status. This data should be retained for a predetermined duration and made available for analysis in the event of an accident or safety concern. According to Regulation (EU) 2019/2144 issued by the European Union, all new vehicles sold in the EU from July 2024 must be equipped with advanced safety systems, including Event Data Recorders (EDRs). These systems must record and store key data related to vehicle operation for accident investigation and vehicle safety improvement. The Centre for Connected and Autonomous Vehicles (CCAV) in the UK proposes in its "Code of Practice for Automated Vehicle Trialing" that EDRs in autonomous vehicles should capture at least 16 types of information at a minimum frequency of 10 Hz, including data on the vehicle's trial mode, location, and speed, as well as actions taken by the driver or safety operator and road conditions.

Vehicle data, a crucial component of contemporary intelligent transportation systems, transcends its primary function of monitoring vehicle operational status. Indeed, through meticulous analysis of vehicle data, the used car market can gain a comprehensive understanding of a vehicle's historical usage, encompassing mileage, maintenance records, and accident history. This establishes a transparent and objective transactional framework, thereby promoting market fairness and efficiency for both buyers and sellers. Furthermore, the role of vehicle data in reflecting component operational status cannot be disregarded. For insurance companies, this data is paramount for assessing vehicle risk and tailoring personalized insurance policies. Concurrently, after-sales maintenance departments can forecast potential failures based on real-time data, formulate maintenance plans in advance, and mitigate customer dissatisfaction and cost escalation stemming from unforeseen breakdowns. At a deeper level, vehicle data unveils driver behavior patterns and in-vehicle system preferences, furnishing valuable resources for advertisers, content

[5] UN Regulation No. 157—Automated Lane Keeping Systems, ALKS, available at https://unece.org/transport/documents/2021/03/standards/un-regulation-no-157-automated-lane-keeping-systems-alks

providers, and even manufacturers themselves to construct user profiles. This enables precision marketing, product optimization, and service innovation.

However, the improper utilization of vehicle data can engender intense market competition disputes, elevate the risk of data leakage, and subsequently infringe upon privacy and property rights. For instance, certain "smart parking" systems possess vulnerabilities that facilitate unverified access to vehicle parking information, potentially exposing sensitive personal data. Autonomous vehicles comprise multiple decision-making and service support systems involving various operators who collect and utilize data independently, which may result in data ownership disputes. Furthermore, vehicle owners may assert the right to copy or transfer vehicle data, potentially conflicting with the interests of autonomous vehicle system operators.

Data of Road Environment

During its operation, a connected vehicle collects road environmental data from its surroundings and interacts with transportation facility-related data to facilitate safe and efficient driving. Firstly, millimeter-wave radar and lidar data are employed to emit and receive signals, measuring parameters like obstacle information, distance, and speed in the surrounding environment. This provides vehicles with high-precision environmental perception capabilities. Secondly, high-definition cameras capture camera data, enabling vehicles to visually perceive traffic signs, moving vehicles, pedestrians, road markings, and other road-related information. Thirdly, high-definition map data is created through 3D modeling of the travel route using onboard lidar, cameras, and other devices. Lastly, traffic facility-related data, such as traffic signal statuses, intelligent traffic signals, and roadside camera and radar data, are essential for autonomous vehicles to adhere to traffic regulations and ensure safety by accessing this information in real time.

Autonomous driving sensors gather personal data of individuals outside the vehicle, such as facial images of pedestrians from videos and license plate numbers of other vehicles. A secure approach involves processing this data within the vehicle, utilizing edge computing technologies to reduce reliance on external servers and prevent the transmission of identifiable personal data externally. Article 6 of China's "Several Provisions on the Administration of Automobile Data Security (Trial)" promotes the lawful, reasonable, and effective use of automobile data, emphasizing the importance of in-vehicle processing and discouraging external sharing unless absolutely necessary. In situations where obtaining personal consent for collecting and sharing personal information outside the vehicle is impractical but essential for ensuring driving safety, anonymization techniques should be applied. This may involve deleting images that can identify individuals or applying partial contour processing to facial information in the images. However, in instances involving vehicle malfunctions, real-time navigation, remote diagnostics, and legal compliance, the remote transmission of road environment data, including personal information, may be unavoidable when obtaining personal consent is not feasible.

The sensors in connected vehicles gather a wide range of environmental data along the vehicle's route, which may involve sensitive locations of official institutions, government personnel, and sensitive activities. One key concern is the regulation of surveying and mapping qualifications for creating high-precision maps. In certain countries, such as China, regulations restrict geographical surveying and data collection to licensed organizations to prevent the leakage of geospatial data that could be exploited by hostile nations for military purposes. Another issue involves data related to government personnel and sensitive activities, as the real-time collection of road environment data by vehicles could lead to the transmission of national security-related data overseas or for military intelligence purposes. For example, Tesla's "Sentry Mode" data collection method attracted scrutiny from Chinese regulators, leading to the introduction of a user-selectable option to disable it in specific areas. This response addresses concerns and offers users a choice in the matter.

Personal Data Rights in Connected Vehicles

Identifiability and Categories of Personal Data

Most data generated by connected vehicles pertain to identifiable natural persons such as drivers, passengers, vehicle owners, and other road users, potentially constituting personal data subject to legal constraints. This data includes directly identifiable information like mobile phone numbers, ID card numbers, and driver's license numbers of drivers and passengers,[6]

[6] Standardization Administration of China. Information Security Technology—Guide for Evaluating the Effectiveness of Personal Information De-identification (GB/T 42460-2023). Standard Press of China, 2023.

as well as indirectly identifiable data such as trip details, vehicle usage data (e.g., data related to driving patterns or distances traveled), or vehicle technical data (e.g., data on vehicle component wear). This data, when cross-referenced with other documents, particularly the Vehicle Identification Number (VIN), may link back to natural persons.

Whether the VIN constitutes personal information has become a focal point of contention across various jurisdictions. In the Gesamtverband case (C-319/22), the European Court of Justice held that the VIN alone does not identify a natural person and is not considered "identified" direct personal information. However, if someone possesses reasonable means to link VIN information to a specific natural person, such VIN information may qualify as personal data and indirectly constitute the personal data of the manufacturer providing the VIN information.[7] Some opinions suggest that the court has created a concept of "indirect personal data" here. In the case of Mr. Yu vs. Chaboshi App (2021) in China, the Guangzhou Internet Court ruled that the vehicle frame number, displayed on the exterior of the vehicle, is merely a code for identifying a specific vehicle and cannot identify a specific natural person. Basic vehicle driving data only includes mileage information, and maintenance and collision data do not reveal the location details of maintenance and repair facilities or specific dates of service, thereby preventing the identification of individuals' whereabouts. The historical vehicle condition information in question solely pertains to the usage of the queried vehicle and does not contain details about specific individuals or assess their behavior or status. As a result, it is not possible to link this information to specific individuals, such as vehicle owners.

Numerous apps related to automotive operations also collect personal data. To clarify the scope of personal data that these apps must collect to fulfill their basic functions, Chinese regulatory authorities issued the "Regulations on the Scope of Necessary Personal Information for Common Types of Mobile Internet Applications" in 2021. It specifies that the necessary personal information for map navigation apps includes location information, departure point, and destination. For online ride-hailing apps, necessary personal information includes the registered user's mobile phone number, passenger's departure point, destination, location information, travel trajectory, and payment information such as payment

[7] European Court of Justice. Gesamtverband der Deutschen Versicherungswirtschaft eV (C-319/22). 2023.

time, amount, and channel (for online ride-hailing services). For car transaction apps, necessary personal information includes the registered user's mobile phone number, purchaser's name, type and number of identification documents, seller's name, type and number of identification documents, vehicle license plate number, and VIN. The authorities aim to instruct these app operators to refrain from collecting any additional unnecessary personal information by specifying particular types of data.

Individual Rights in Personal Data Protection

The GDPR establishes a comprehensive framework for protecting personal data rights in Europe, with many of its principles also being incorporated into China's Personal Information Protection Law. Under this legal framework, individuals, as data subjects, are entitled to a series of rights by law, including the right to give explicit consent, refuse, or withdraw consent, the right to copy and transfer personal data (also known as the right to data portability), the right to be informed of the specifics of data processing activities, the right to have personal data deleted, and the right to remedies such as compensation in case of infringement.

To ensure the enforcement of these personal data rights, controllers and processors of vehicle data must integrate essential tools into their information systems to facilitate the exercise of these rights. The EDPB recommends that vehicle and device manufacturers, service providers, and other data controllers implement a personal data management system (profile management system) to store known driver preferences and facilitate easy modification of personal privacy settings at any time. This system should centrally manage each data setting for every data processing instance, specifically allowing data subjects to access, delete, erase, and export their personal data from the vehicle system upon request. Drivers should have the ability to stop the collection of certain types of data temporarily or permanently at any time, unless the controller has a specific legal justification for continuing to collect that particular data.[8]

China's National Information Security Standardization Committee proposes that convenient functions for personal information management, such as access, copying, and deletion, should be provided to personal information subjects. When products or services involve interactive

[8] European Data Protection Board. Guidelines 01/2020 on Processing Personal Data in the Context of Connected Vehicles and Mobility Related Applications. 2020.

operations, such as websites, in-vehicle applications, or mobile applications, the personal information management functions should be interactive, with easily accessible entry points prominently displayed for individuals to manage their personal information.

A unique challenge related to personal data in autonomous vehicles is determining if individuals in the road environment have corresponding data rights. The cameras installed on autonomous vehicles inevitably capture large amounts of personal information during operation, even though this information is not the primary focus of processing. A practical solution is to desensitize these data through anonymization before remote transmission. Similarly, Google Street View automatically blurs license plate numbers and faces in images to prevent personal identification; users can also request the removal or further blurring of specific images through Google's reporting tool, applicable to scenarios where individuals, households, or vehicles may be identifiable. However, in cases where general AI collects massive amounts of road environment data for training purposes, achieving complete anonymization of personal information in the road environment is nearly impossible, if not economically unsustainable. Therefore, we suggest referring to Article 11 of GDPR, which allows exemptions from consent or the right to object and erasure when personal data is inevitably collected without the intention of identifying individuals but additional personal information must be collected to fulfill obligations such as consent and notification. As a pragmatic approach, it is unrealistic to strictly apply all personal information rights regulations in autonomous driving scenarios.

Legal Obligations to Data Processors

In the realm of autonomous vehicles, the key entity accountable for the processing of personal data is the personal information controller. Their obligations can be classified into two main types: the first involves supportive duties to address the rights of the personal data subjects, while the second entails specific security management obligations essential for maintaining the personal data in a secure manner. It should be noted that under Chinese law, the personal data processor is equivalent to the 'controller' under the GDPR, whereas the GDPR's 'processor' is considered as the 'entrusted processing' in the Chinese legal framework.

With respect to supportive obligations, Chinese national standards have further elaborated on the specific requirements for related safeguard measures. For instance, (1) To ensure the deletion of personal information within ten working days of receiving a request, it is essential to set up a structured directory of personal information, thereby achieving traceable management of personal information. (2) When seeking personal consent, it should be communicated to the individual through at least one prominent mean. Such means may include a separate section in the user manual, voice announcements, pop-up notifications on the vehicle display panel, interaction through vehicle-related applications, a separate section in the car sales agreement, a separate section in the maintenance service agreement, or interaction through travel service applications, among others. (3) When informing data subjects about the retention periods for different types of personal information, the details should be specific and clear, such as 30 days or one year, etc.

In terms of security management obligations, Chinese national standards have introduced additional requirements in line with the principles of minimal necessity and the privacy impact assessment system. For example, (1) Once the specific purpose has been fulfilled, the vehicle data processor should promptly delete the relevant personal data or anonymize it. (2) The vehicle data processor should conduct a risk assessment of vehicle data, which generally includes the identification of automotive data, the identification of data processing activities, the identification and evaluation of vehicle data security risks, and may be conducted through self-assessment or third-party assessment. (3) The individual responsible for vehicle data security management should be appointed by the primary person in charge or the data security officer of the automotive data processor and should be familiar with Chinese data security and personal information protection policies and regulations, as well as relevant experience in security management. (4) The vehicle data processor should establish a comprehensive security emergency response mechanism, conduct at least one emergency drill per year, and preferably support forensic analysis after a security incident through mechanisms such as vehicle data certification and traceability. (5) The vehicle data processor should accept complaints related to vehicle data security through means such as telephone or instant messaging platforms, generally process the complaints within ten working days of receipt, and maintain a complete record of the process and outcome of the handling.

Violations of these obligations by vehicle data processors may incur civil liability for infringement and even criminal sanctions. For instance, the Beijing Internet Court adjudicated a civil dispute case where an APP operator providing automotive information services unlawfully provided its users' mobile numbers to car brand dealerships, leading to the users receiving numerous quotation calls from nine local 4S stores of three brands, and was found to have infringed on the rights to personal information. In 2023, the police department in Guangdong investigated a criminal case where an automotive service company, to expand its car repair brokerage business, colluded with insurance companies, towing companies, and road administration department staff to illegally obtain the information of car accident owners they came across during their work, misused this personal information to induce accident owners to repair their vehicles at designated locations by offering benefits such as maintenance and fuel cards, earning "brokerage" fees from the information, and ultimately arrested 47 suspects with the case involving over 3 million RMB in transaction volume.

Property Rights of Vehicle Data

Value of Data Property Rights

Autonomous vehicles generate data while in operation, and the advancement of autonomous driving technology requires the collection or sharing of this data. Industries such as insurance, finance, sales, research and development, and maintenance are also increasingly interested in obtaining vehicle data. Aligning with the structure of the vehicle data value chain, existing research has proposed eight types of business models for monetizing vehicle data.[9] According to a McKinsey report, by 2030, shared mobility and automotive data-related services are expected to become consistent and growing revenue streams, apart from traditional vehicle sales revenue. The commercialization of vehicle data could potentially generate a total revenue of 450 billion to 750 billion.[10] In terms of application methods, the monetization of vehicle data can be divided into two categories: direct monetization and indirect monetization. Direct monetization refers to the

[9] Christian Koehler, et al. Monetizing automotive data —A perspective on how data drives revenues and growth in future automotive business models, January 2020, available at: https://www.strategyengineers.com/-/monetizing_automotive_data

[10] C. Martens and T. Schneiderbauer, Unlocking the full life-cycle value from connected-car data, McKinsey &Company, pp. 1–15, 2021, available at https://www.mckinsey.de/industries/automotive-and-assembly/our-insights/unlocking-the-full-life-cycle-value-from-connected-car-data

creation of new revenue streams by offering data or insights to customers or partners, a practice commonly referred to as "external" monetization. Indirect monetization involves enhancing business performance through data-driven insights or services, encompassing a more extensive form of "internal" monetization.[11] The value of vehicle data can be illustrated from three aspects.

Vehicle data is of paramount importance in the research and development of autonomous driving technology. Simulation testing is an indispensable component of the development and validation phases of autonomous driving systems. By constructing virtual traffic environments and simulating various complex driving scenarios, data from simulation tests is collected to evaluate the performance of autonomous driving systems. This data facilitates the early identification and optimization of potential issues. Furthermore, analyzing accident data provided by automotive repair service providers enables vehicle manufacturers to identify potential system flaws and implement corresponding improvements. In recent years, many research institutions and enterprises have open-sourced large-scale autonomous driving datasets, such as the Waymo Open Dataset, nuScenes, and Argoverse. These datasets encompass diverse driving scenarios and annotation information, serving as valuable resources for autonomous driving technology developers. In cities like Beijing and Shanghai, specialized agencies, as well as alliances, have launched Vehicle Data Pool for intelligent and connected vehicles, unifying data standards, aggregating data collected from multiple sources, and circulating and utilizing it for free or at a cost, developing a range of diversified data service solutions.

Vehicle data holds substantial commercial value for third-party service providers, enabling them to execute targeted marketing campaigns with greater precision. Insurance agencies can utilize historical accident records, driving behavior data (such as frequent acceleration and braking), and vehicle maintenance records to more accurately assess the risk level of car owners, thereby formulating personalized insurance pricing strategies. Compared to traditional pricing methods based on age, gender, and occupation, this big data-based pricing is more scientific and precise, enhancing customer purchase intention and satisfaction. Some in-vehicle apps can recommend suitable maintenance services, accessory products, travel

[11] Amadeus Tunis, Car data monetisation needs a reality check, June 2022, available at: https://www.automotiveworld.com/articles/car-data-monetisation-needs-a-reality-check/

routes, etc., based on user vehicle information and behavior data, achieving precise marketing. These personalized recommendations greatly enhance user purchase intent and conversion rates. Sales providers can suggest appropriate accessory products and promotional information tailored to individual car owners' needs and vehicle conditions, enabling precise marketing. This data assists sales providers in accurately forecasting the demand for various accessories, optimizing inventory management, and preventing both overstocking and out-of-stock scenarios.

Vehicle data also holds social value by optimizing the allocation of public service resources. The mining and analysis of vehicle data have facilitated evidence-based decision-making in road inspection, maintenance planning, and the formulation of related policies (Schellekens, 2022). Hangzhou's City Brain parking system platform has accessed a large amount of parking lot and berth data, aggregating massive data covering all elements of the parking ecosystem. By analyzing real-time parking data, Hangzhou has installed parking guidance screens and smart parking guidance signs in multiple key areas of the city (such as hospitals, schools, business districts, etc.), real-time releasing information on available parking spaces in nearby parking lots, guiding car owners to quickly find parking spaces, and effectively alleviating the difficulty of parking. Utilizing real-time traffic flow data, smart traffic signal systems can automatically adjust signal timing, optimize traffic flow, and alleviate congestion. Furthermore, by analyzing vehicle travel trajectories and hot spot area data, city planning departments can more accurately predict future traffic flow changes, scientifically plan road networks, public transportation lines, and transportation hubs, and enhance the overall efficiency of urban transportation systems.

Ownership of Vehicle Data

The property value of data has gained increasing recognition, earning it the metaphorical title of 'oil' in the information age. In China, data has been officially designated as the fifth major factor of production at the national level, alongside land, labor, capital, and technology. Consequently, the establishment of the National Data Administration aims to facilitate data circulation and reuse. However, whether data, as a form of property, possesses "ownership" is a global theoretical challenge. Europe introduced the concept of "data ownership" in draft official documents starting from 2015, but ultimately adopted "data access rights" in the passed Data Act, merely confirming that data sources have the right to obtain data from data controllers, without specifying the rights held by data controllers over the data they possess. China's 'Twenty Measures for Data' outline plans to establish a segmented property rights operation mechanism, including data resource holding rights, data processing and use rights, and data product operation rights, aiming to directly address the challenges of data property rights allocation.[12] The ownership of vehicle data can be divided into three situations for separate handling.

The first scenario pertains to data that is collected and generated independently, solely belonging to one party. For example, autonomous driving research and development enterprises use their own equipment to collect original natural environment data or synthesize virtual data using technical methods. This data does not infringe upon the interests of any other party, nor is there a valid reason for others to contest its ownership. In cases where there are no disputes regarding the ownership of such data, it rightfully belongs to the data controller who collected and generated it.

The second scenario involves multiple data contributors collaborating to build a shared data resource pool. The value of data needs to reach a certain scale to manifest. In the autonomous driving industry, there are numerous data sharing platforms where different entities contribute their data and have access rights to the collective data pool. Such collaboratively created data pools should be handled with reference to the rules of "common" property rights, indicating that all participating parties jointly own the data pool. The ownership structure, whether shared or joint ownership, should be established through contractual agreements or industry norms agreed upon by the data pool participants. In the absence of clear contractual agreements or industry standards, shared ownership should be presumed, ensuring that each contributing party has equal property rights over the data pool.

The third scenario involves data collected by data controllers from external organizations, individuals, or property owned by others, raising issues regarding the allocation of rights between data sources and data controllers. The autonomous driving ecosystem contains numerous IoT devices, and the system controllers of these IoT devices collect a large amount of data, which often originates from holders of various IoT

[12] "Central Committee of the Communist Party of China and State Council. (December 2022). 'Guiding Opinions on Building a Data Foundational System to Better Utilize the Role of Data Elements.'"available at https://www.gov.cn/zhengce/2022-12/19/content_5732695.htm

devices. For instance, original equipment manufacturers, data integrators, infrastructure operators, and car owners all play significant roles in the generation and utilization of vehicle data. Car owners typically view themselves as the rightful owners of data collected by sensors and other components within their vehicles. Original equipment manufacturers, while acknowledging car owners or leaseholders as the owners of connected vehicle data, may access and manage the data through user agreements. Data integrators perceive themselves as the owners of the data they distribute, as it is derived from vehicle data. Infrastructure operators consider themselves as the owners of road-related data provided and exchanged by their infrastructure systems.[13]

In cases of conflicts over property rights, an ideal approach is to establish rules for allocating data property rights through contracts for data controllers and data sources. However, reaching a fair agreement on data ownership among vehicle data participants is currently challenging, often resulting in the absence of contractual clauses addressing this issue. Therefore, resolving this matter will require enhancements to civil legal regulations in the future. The specific solution should adopt the theory of "parallel ownership," fully reflecting the non-exclusive and replicable characteristics of data property rights to fairly and fully leverage the reuse value of data. Under this framework, if the data controller acquires data through legitimate means without violating mandatory legal requirements or infringing on rights such as trade secrets and personal data, they can claim ownership of the data. Simultaneously, data initiators have the right to request copies and transfers of the data they have contributed, thereby upholding their data rights. In summary, data controllers and data initiators should each hold ownership of the data, contingent upon respecting the legal rights of the other party.

Data Reuse under the Principle of Fairness

Automobile data can be reused across different entities and application scenarios, thereby generating greater economic and social value. Nevertheless, certain automakers have established a data control framework

[13][Johanna Zmud, et al. Data Ownership Issues in a Connected Car Environment: Implications for State and Local Agencies, November 2016, available at: https://tti.tamu.edu/tti-publication/data-ownership-issues-in-a-connected-car-environment-implications-for-state-and-local-agencies/]

granting exclusive access rights to uphold their dominant position in the sales and post-sales service of their branded vehicles.[14] In Europe, one of the objectives of The Data Act is that manufacturers and designers of smart products will have an obligation to share data with the users of those products and with third parties designated by users. In 2022, the European Union specifically issued an initiative titled "Access to Vehicle Data, Functions, and Resources," which elaborated on the necessity of promoting the reuse of vehicle data. China's "Twenty Measures for Data" proposes to guide leading industry enterprises and internet platform enterprises to play a leading role, promote two-way fair authorization with small and medium-sized enterprises (SMEs), jointly and rationally use data, and empower the digital transformation of SMEs. When it comes to ensuring fair access and utilization of automotive data, the key consideration is determining the equitable rules that data controllers with substantial automobile data holdings must adhere to when sharing data externally. Firstly, data should be open to SMEs and individuals according to the principle of fairness; secondly, data should be made available to organizations with public service attributes, such as governments and research institutions, according to the principle of fairness.

The primary targets for automobile data controllers to open data to enterprises and individuals are enterprises that hold a dominant market position or possess non-substitutable data resources within the autonomous driving ecosystem. One approach is to handle data requests in accordance with the principle of "providing as much as possible." Otherwise, the reasons for being unable to provide the data externally must be explained. That is, these automobile data controllers have a mandatory contractual obligation and cannot refuse requests for corresponding data without providing sufficient reasons. Data controllers may decline requests for personal data due to cybersecurity concerns related to the personal information contained within the data. Additionally, they may require data applicants to pay the applicable fees. Secondly, a reliable data circulation infrastructure and unified technical standards are essential guarantees to enhance the quality of data circulation and utilization, while simultaneously striking a balance between data circulation and utilization and ensuring security protection. The "external server" and the "on-board application platform" constitute viable technological solutions (Kerber,

[14] Bertin Martens & Frank Mueller-Langer, Access to digital car data and competition in aftersales services, JRC Digital Economy Working Paper 2018–06, European Union, 2018.

2018). For example, both Chinese and Western data regulations include provisions on the portability of personal data, and the implementation of this system highly relies on technical standards for data circulation. Thirdly, it is imperative to define the scope of "non-substitutable data resources," encompassing infrastructure data that cannot be acquired through alternative channels and data that is excessively costly to obtain through alternative channels, rendering it commercially infeasible.

The primary concern regarding the opening of automobile data controllers to the government pertains to the methodology employed by the government to acquire enterprise data. First and foremost, the government may mandate enterprises to collaborate in providing data for statistical, law enforcement, security, and other administrative functions. Acquiring enterprise data through this method necessitates legislative authorization. For instance, autonomous driving regulations in numerous regions mandate enterprises to promptly report data on vehicle safety incidents to transportation authorities. Secondly, the government may acquire data from enterprises to optimize public services and enhance efficiency. Data belongs to enterprises, and the government can only acquire enterprise data through market mechanisms without legal authorization. Thirdly, establish a national data infrastructure to promote data sharing and utilization. For example, in the Mobile Data Act(Draft) released in May 2024, Germany proposed to provide and use mobility data through the National Access Point to support innovation and efficiency in modern transportation services.[15]

Rules for Cross-Border Data Flow Rules of Connected Vehicles

International Data Competition and Data Distrust among Nations

Regulations governing cross-border data flows have emerged as a pivotal concern in contemporary digital economic policies and international political dynamics. In April and June 2024, respectively, China and Germany signed the "Joint Intention Statement on Cooperation in the

[15] "Referentenentwurf: Entwurf eines Gesetzes zur Bereitstellung und Nutzung von Mobilitätsdaten und zur Änderung des Personenbeförderungsgesetzes." Bundesministerium für Digitales und Verkehr, Bearbeitungsstand: 07.05.2024.

Field of Automated and Connected Driving" and the "Memorandum of Understanding on Sino-German Cooperation on Cross-Border Data Flows." On July 31, 2024, the White House, along with over ten countries and regions including the European Union, held a special meeting on vehicle data security, aiming to unify cross-border data flow rules to address national security risks associated with connected vehicles. In a request for comments issued in March 2024, the U.S. Department of Commerce claimed that China's legal requirements for connected vehicles to provide data to the government threaten individual privacy and U.S. national security, recommending that the government take specific actions to restrict cooperation with China in the design, development, manufacturing, or supply of connected vehicles within the United States. According to the plan outlined in the notice, the U.S. Department of Commerce intends to issue proposed rules specifically targeting Chinese connected vehicles, using the pretext of national security risks to American data. Overall, the regulations governing cross-border data transfers are not solely legal matters; they are also influenced by economic considerations related to data competition and political considerations concerning national security. This reflects a lack of mutual trust among nations.

The fundamental requirement for cross-border transfers of personal data within the automotive industry is the implementation of standardized contractual clauses. A survey conducted by the European Automobile Manufacturers' Association and related institutions in 2020 indicated that nearly 85% of companies use standard contracts as the compliance framework for data exports, and only 9% of companies do not transfer data outside the European Union.[16] Before the introduction of new regulations on cross-border data flows in March 2024, the primary method for the export of personal data in the Chinese automotive sector was through the "security assessment" organized by the Central Cyberspace Affairs Commission. This administrative procedure was considered costly and inefficient, not well-suited to the basic principles of data flows. After the new regulations were issued and implemented in March 2024, for the export of personal information of over 100,000 but less than one million individuals (excluding sensitive personal information) or sensitive personal

[16] "Schrems II: Impact Survey Report(November 2020)," Business Europe, DIGITALEUROPE, the European Round Table for Industry, and European Automobile Manufacturers Association, available at https://www.digitaleurope.org/resources/schrems-ii-impact-survey-report/

information of less than 10,000 individuals, a standard contract for the export of personal information can be established with the overseas recipient or data export can be conducted through personal information protection certification. Considering the sales distribution of foreign new energy vehicle brands in China, most of them fall within the range of over 100,000 to less than one million vehicles. For example, Tesla's vehicle sales in China exceed 210,000 units. Compared to "certification," "standard contracts" are a more popular method for cross-border data transfers, and it is expected that "standard contracts" will become the primary pathway for data exports.

Data localization has emerged as a crucial strategy adopted by certain automotive manufacturers to mitigate risks associated with global political dynamics, particularly the risks associated with alterations in political relations between China, the United States, and Europe. The automobile manufacturing and sales supply chain is a highly globalized market. The optimal international economic order entails strategically locating manufacturing facilities in regions with the most suitable labor force and established industrial infrastructure. Similarly, research and development centers should be established in areas where talent and technological expertise are abundant. Additionally, data centers should be situated in locations that offer favorable land prices, favorable climatic conditions, and cost-effective electricity supply. However, this ideal environment is subject to international political influences. The existence of cross-border data flow regulations results in frequent regulatory oversight of daily data flows. One solution is to deploy data locally to avoid cross-border data flows. In 2021, Reuters reported that automobile manufacturers such as BMW, Daimler, and Ford had already set up data centers in China.[17] On August 14, 2023, addressing the hotly debated "Sentry Mode," Tesla posted on Weibo that the company had established a data center in China and that all data generated by vehicles sold in the Chinese mainland market would be stored within China.

[17] "Exclusive: China Plans New Rules for Global Automakers to Move, Store Car Data Locally." Reuters, Thomson Reuters, 27 May 2021, available at www.reuters.com/business/exclusive-china-plans-new-rules-global-automakers-move-store-car-data-locally-2021-05-27/

Building a Trusted Framework

The policy on cross-border data flows essentially seeks to strike a balance between "security assurance" and "facilitating development." A credible framework is beneficial to all countries. The primary concerns of nations regarding vehicle data pertain to judicial sovereignty, national security, and personal information protection. In the realm of judicial sovereignty, China has consistently opposed the expansive jurisdiction of the United States, which entails accessing data within China without explicit authorization. Additionally, China expresses concerns regarding the potential erosion of its judicial sovereignty when data is stored abroad without the provision of domestic backups. Regarding national security, the key concerns center around automotive companies reporting sensitive data to governments and organized national cybersecurity forces monitoring and stealing data, which is a crucial issue leading to mutual complaints and protests among countries. In terms of personal information protection, the frequent occurrence of vehicle data breaches poses a significant challenge. This challenge is particularly acute when personal information infringements occur abroad, hindering citizens from asserting their rights. In 2023 alone, there were at least over 20 major data breaches related to automotive companies worldwide, including Toyota leaking vehicle equipment identification numbers and vehicle locations in Japan, and Tesla leaking a large amount of employee-related records due to "misconduct by internal employees." Without a coordinated international protection mechanism, countries' own personal information protection mechanisms will be susceptible to vulnerabilities.

Establishing a mutually trustworthy framework for cross-border data flows is an inevitable trend and the ultimate choice. The WTO's e-commerce directive has already included rules on cross-border data flows as important content for negotiation, and many countries have achieved cross-border free flow of data in their economic agreements. Additionally, organizations such as DECP, CPTPP, APEC, and G7 have proposed frameworks for mutual recognition of cross-border data. However, these frameworks encounter challenges in their effective implementation, compelling the adoption of more practical solutions within the vehicle data industry. First and foremost, it is imperative to strictly control the scope of sensitive data to minimize unnecessary regulatory burdens. These strictly regulated data should primarily encompass audio and video recordings of individuals within vehicles, precise driving trajectory data, and data pertaining to sensitive locations such as military and government facilities that are not accessible from vehicles. Countries should establish a fundamental consensus on this negative list. Secondly, there is a need to establish a mechanism for mutual sharing and collaborative handling of data risks. Illegal data transactions on the dark web and hacker attacks that steal data should become targets for joint combat by all countries. This requires the establishment of a global vehicle data risk sharing and judicial assistance mechanism to form an efficient and convenient governance structure for vehicle data security. Thirdly, establishing technical standards for vehicle data sharing is crucial to promote interoperability and facilitate the reuse of vehicle data across countries. For example, technical standards can be leveraged to facilitate the trusted flow of data within the framework of the World Forum for Harmonization of Vehicle Regulations (WP.29) under the United Nations Economic Commission for Europe (UNECE).

It is noteworthy that certain free trade zones in China are actively promoting cross-border data flows of vehicles through a classified management approach. For instance, the Lingang New Area of the China (Shanghai) Pilot Free Trade Zone released the "Scenario-based Cross-border Data in the Field of Intelligent and Connected Vehicles (General Data List)" in 2024, outlining measures for real-time and convenient handling of non-sensitive data. Although case-by-case efforts may provide temporary solutions, they cannot be a universal approach. Consequently, addressing the issue requires a comprehensive solution within the framework of international law.

Conclusion

The value and risks associated with vehicle data vary significantly across different scenarios, being highly dependent on the governing framework. Highly structured data serves as the "oil" for the operation of autonomous vehicles, while fragmented and incompatible data can only become an expensive "liability" in automotive operations. Vehicle data with inadequate cybersecurity governance mechanisms may turn into a "powder keg" that could potentially lead to the collapse of enterprises. Organizations and individuals holding vehicle data must clarify its specific connotations and subdivisions, identify its value in various application scenarios, and transform raw "data resources" into refined "data assets." For national institutions entrusted with data governance responsibilities, it is crucial to continuously translate abstract security concepts into tangible practices, avoid institutional costs arising from uncertainties in data compliance risks, and create a favorable legal environment for the development of the autonomous vehicle industry. From the perspective of the global vehicle data governance framework, we need a mutually trustworthy and coordinated international environment to enhance the level of vehicle data governance in accordance with the development patterns of the globalization of the automotive industry, thereby promoting free competition among relevant entities.

References

Kerber, W. (2018). Data governance in connected cars: The problem of access to in-vehicle data. *Journal of Intellectual Property, Information Technology and E-Commerce Law, 2018*, 310–331.

Schellekens, M. (2022). Data from connected cars for the public cause. *Computer Law & Security Review, 45*.

PART IV

A Survey Findings Report in Hong Kong and Shenzhen

CHAPTER 9

Chinese People's Preferences for Automated Vehicle Designs: A Survey Findings Report in Hong Kong and Shenzhen

Shiling Xiao, Chunyan Ding, and Ruiping Fan

Introduction

It is not imprudent to claim that the history of humanity is a history of using and developing tools and machines. The advent of machines and technological tools has played a crucial role in the progress of human society. In the twenty-first century, we are entering an age in which machines

S. Xiao (✉)
Law Department, The University of Manchester, Manchester, UK
e-mail: shiling.xiao@manchester.ac.uk

C. Ding
School of Law, City University of Hong Kong, Kowloon, Hong Kong
e-mail: chunding@cityu.edu.hk

R. Fan
Department of Public and International Affairs, City University of Hong Kong, Kowloon, Hong Kong
e-mail: safan@cityu.edu.hk

C. Ding, R. Fan (eds.), *Chinese Ethical and Legal Perspectives on Automated Vehicles*,
https://doi.org/10.1007/978-981-96-1277-2_9

not only promote human welfare but also determine the allocation of inevitable harms that they cause. Automated vehicles (AVs) are such machines. Undoubtedly, AVs have the potential to enhance road safety, reduce traffic congestion, and provide increased mobility for individuals who face challenges with traditional transportation methods. However, in ethical dilemmas and legal puzzles faced by AVs, envision a situation where an AV must decide whether to swerve, potentially endangering three pedestrians crossing the road or maintain its course, risking harm to a group of schoolchildren inside the vehicle. This scenario underscores the complex ethical and legal choices AVs must confront, involving the allocation of risk and harm among different stakeholders on the road. Car manufacturers, system designers, legislators, and law-enforcement authorities are currently grappling with these moral and legal dilemmas. Solutions to these moral and legal puzzles can hardly be a single and universal doctrine that is applicable to every culture and legal system.

A significant number of philosophical and legal research concerning the intractable issues of AVs has been published. However, it has generally been examined and discussed from Western perspectives (Carlsson et al., 2010; Bonnefon et al., 2016; Awad et al., 2018). For example, the Moral Machine, an online experimental platform designed by a group of international researchers led by MIT to explore the ethical dilemmas faced by AVs, is arguably the most up-to-date and comprehensive survey, collecting approximately 40 million decisions in ten languages from 233 countries and areas. However, as the survey report shows, most decisions were drawn from European and American states and areas. The participation of Chinese respondents is not adequately representative (Awad et al., 2018). In China, there is no survey to find out how Chinese people respond to ethical and legal dilemmas relating to AVs. Our survey findings presented below fulfill this gap.

What are the preferences of Chinese people when facing these ethical dilemmas and legal puzzles? Are Chinese people's choices different from the choices of people from other parts of the world? The answers to these questions might not only interest Chinese people but also arouse curiosities beyond China. Moreover, Chinese people's preference for AV designs may also have a significant implication for car manufacturers, system designers, legislators, and law-enforcement authorities.

First, our survey can inform ethical considerations. Surveys provide valuable insights into public attitudes, beliefs, and values regarding ethical dilemmas associated with AVs (Othman, 2021). By surveying individuals in two representative cities—Hong Kong and Shenzhen, we gain insights into societal expectations, concerns, and priorities related to AV designs and operations. These insights can inform ethical deliberations by providing a diverse range of viewpoints for policymakers and ethicists to consider. For instance, preferences regarding prioritization in accidents, data privacy concerns, or decision-making algorithms in AVs can be better understood through our survey, helping in the formulation of ethical guidelines or principles.

Second, survey results serve as valuable inputs for policymakers in developing regulatory frameworks for AVs. In general, surveys play a crucial role in informing policy development and regulatory frameworks (Althaus, 2003). By capturing public sentiment and preferences, policymakers can gauge public acceptance, identify potential areas of concern, and understand the impact of various policy options (Cruz et al., 2023). Such information contributes to the development of more inclusive and informed policies and regulations surrounding AVs.

Third, surveys can enhance public engagement. Surveys facilitate public engagement and participation in discussions on AV ethics and regulations. By providing individuals with a platform to express their opinions and concerns, surveys foster a sense of inclusion and democratic decision-making (Nabatchi & Amsler, 2014). They allow the public to contribute to shaping ethical guidelines and legal frameworks, promoting transparency and accountability in the development and implementation of AV technologies.

Fourth, this survey collects people's opinions on critical legal rules surrounding AVs, including the allocation of liability after accidents, the protection of privacy and personal data, and the allocation of the burden of proof in a dispute, which could help the legislators to lay down laws that regulate them. Legal rules profoundly impact people's behavior and influence the development of this technology and even the whole industry. However, the survey on people's opinion on regulations of AVs is nil. Ours fills the gap.

Last but not least, conducting the survey among Chinese people provides insights into preferences and attitudes that are rooted in the cultural and societal context of China. Chinese cultural values, norms, and perceptions may differ from those in Western societies, impacting preferences regarding AV designs and ethical considerations. By conducting the survey in China, this research enriches the global discourse on AV ethics, policy, and design by incorporating perspectives from a major non-Western market. This global diversity of perspectives is essential for fostering inclusive and equitable development of AV technologies on a global scale. Understanding these cultural nuances is essential for the development and deployment of AVs in China, ensuring that they align with local values and expectations. China represents one of the largest and fastest-growing markets for AV technologies. By focusing on Chinese preferences, the survey addresses the specific needs and interests of this significant market segment. Insights gained from the survey can inform industry strategies and product development tailored to the Chinese market, enhancing competitiveness and market penetration for AV manufacturers and service providers.

The Survey Design

Data for this study were collected through an online survey conducted both in Hong Kong and Shenzhen in February 2023. The surveys were carried out via online panels hosted by a globally recognized data collection company, Kantar. The company's online panels contained an opt-in list of 56,000 individuals relative to the population size included in this study. We successfully recruited 1000 participants online to respond to our questionnaires. Specifically, 500 participants are from the Hong Kong Special Administrative Region, and another 500 participants are from the city of Shenzhen in mainland China.

The selection of Hong Kong and Shenzhen as the target locations for our survey is grounded in the strategic aim of capturing diverse perspectives within the Chinese context. Both cities, while in close proximity geographically, present unique characteristics in terms of culture, economy, legal systems, and traditions, making them valuable representatives of broader Chinese perspectives.

Geographically, Hong Kong and Shenzhen share a border. However, their topographical and environmental features differ significantly. Hong Kong is characterized by its iconic skyline, hilly terrain, and a blend of urban and rural areas, which foster a unique landscape. On the other side of the border, Shenzhen, known for its rapid development, showcases a more expansive and modern urban layout, reflecting the fast-paced economic growth that has defined the city.

Demographically, Hong Kong and Shenzhen exhibit both similarities and differences that further enrich the breadth of perspectives in our survey. In terms of similarities, both cities have experienced significant population growth fueled by migration and urbanization. The diverse influx of individuals from various regions has contributed to a blend of demographics, creating cosmopolitan environments. However, differences emerge in the demographic composition of these two cities. Hong Kong, with its longer history and established international presence, has a more diverse and aging population, reflecting a mix of local, expatriate, and immigrant communities. In contrast, Shenzhen, as a relatively newer city, has a younger population, driven by a surge of migrants from other provinces and cities of China who are attracted by its economic opportunities.

Culturally, Hong Kong and Shenzhen are influenced by both Chinese heritage and global interactions. Hong Kong has a rich history of mixing Eastern and Western cultural elements due to its colonial past. This cultural diversity is manifested in its cuisine, language, and daily practices. Shenzhen, as a comparatively newer city, has experienced rapid modernization, and its culture is influenced by a mix of traditional Chinese values and contemporary trends, especially in technology and innovation.

Economically, Hong Kong and Shenzhen stand out as vital financial and technological hubs, respectively, with different areas of specialization. Hong Kong serves as a global financial center, fostering international trade and commerce. Shenzhen, on the other hand, has become synonymous with innovation and technology, housing numerous tech companies and serving as a major manufacturing hub.

The legal systems in Hong Kong and Shenzhen exhibit significant distinctions. Hong Kong, with its historical ties to British law, follows a common law legal system that is distinct from the mainland Chinese legal framework under the "One Country, Two Systems" principle. Shenzhen, as part of mainland China, adheres to the civil law legal system of Mainland China. This legal divergence introduces an interesting dynamic in understanding how individuals in these two cities approach legal considerations related to AVs.

By selecting participants from both Hong Kong and Shenzhen, our survey aims to explore the nuanced perspectives that arise from the diverse geographical, cultural, economic, legal, and traditional backgrounds of these two cities. This approach allows us to gain insights into whether and how regional variations influence decision-making on ethical and legal aspects of AVs in the Chinese context. Table 9.1 below shows a general overview of the age, gender, education, driving experiences, and minor, pregnant women and elderly family members.

Table 9.1 Descriptive analysis of the respondents in this survey

Variables	*Shenzhen*	*Hong Kong*	*Total*
Age group			
18–29	147 (29.40%)	84 (16.80%)	231 (23.10%)
30–39	148 (19.60%)	97 (19.40%)	245 (24.50%)
40–49	119 (23.80%)	103 (20.60%)	222 (22.20%)
50–59	65 (13%)	114 (22.80%)	179 (17.90%)
60 and above 60	21 (4.20%)	102 (20.40%)	123 (12.30%)
Gender			
Male	284 (56.91%)	238 (47.60%)	522 (52.25%)
Female	215 (43.09%)	262 (52.40%)	477 (47.75%)
Education			
Primary and below	0 (0%)	9 (1.80%)	9 (0.90%)
Secondary school	22 (4.43%)	165 (33%)	187 (18.76%)
Tertiary	397 (79.88%)	274 (54.80%)	671 (67.30%)
Graduate school and above	78 (15.69%)	52 (10.40%)	130 (13.04%)
Driver's licence			
Have	440 (88.00%)	331 (66.20%)	771 (77.10%)
Do not have	60 (12.00%)	169 (33.80%)	229 (22.90%)
Frequency of driving			
Never	3 (0.68%)	29 (8.76%)	32 (4.15%)
Seldom/once per month	37 (8.41%)	79 (23.87%)	116 (15.05%)
Sometimes/2–3 times per month	34 (7.73%)	69 (20.85%)	103 (13.36%)
Often/once a week	60 (13.64%)	76 (22.96%)	136 (17.64%)
Always/A few times a week	167 (37.95%)	53 (16.01%)	220 (28.53%)
Everyday	139 (31.59%)	25 (7.55%)	164 (21.27%)
Pregnant women/elderly in household			
Yes	218 (42.60%)	232 (46.40%)	450 (45%)
No	282 (56.40%)	268 (53.60%)	550 (55%)
Children under 12 in household			
Yes	270 (54%)	143 (28.60%)	413 (41.30%)
No	230 (46%)	357 (71.40%)	587 (58.70%)

General Findings

Before demonstrating the decisions of participants on various ethical and legal dilemmas relating to AVs, we want to highlight that, according to our survey, participants have shown an increasing level of interest as the levels of driving automation increased. The level of interest in Level 1 AVs was 2.78 (out of 4), Level 2 was 2.8, Level 3 was 2.92, Level 4 was 2.95, and Level 5 was 3.05. When comparing the level of interest in the five levels of AVs between the two cities, participants in Shenzhen have shown a significantly higher level of interest in Levels 1 to 3 compared to participants in Hong Kong. There was no significant difference in the level of

Table 9.2 People's attitude towards the benefits brought by Avs between two cities

Item	*Shenzhen*	*Hong Kong*	*Variation*
Fewer crashes	3.17	2.75	0.41***
Reduced severity of crashes	3.16	2.79	0.37***
Improved emergency	3.13	2.78	0.34***
Response to crashes	3.23	2.87	0.36***
Less traffic congestion	2.94	2.65	0.30***
Shorter travel time	3.00	2.69	0.31***
Lower vehicle emissions	2.91	2.64	0.27***
Better fuel economy	3.00	2.78	0.22***
Lower insurance rates	2.82	2.48	0.34***

*** $p < 0.001$

interest in Levels 4 and 5 between the two cities. Participants have also shown interest in driving or riding AVs. Between the two cities, however, participants from Hong Kong showed a lower level of interest compared to participants in Shenzhen, across all levels from Level 1 to Level 5.

Our survey has also shown that participants from both Shenzhen and Hong Kong have confidence in AVs and believe that AVs will bring various benefits to the users and society, as shown in Table 9.2 above. Nonetheless, participants in Hong Kong showed a lower level of optimism for all nine benefits compared to participants in Shenzhen.

The survey questions were categorized into two sections: ethical aspects and legal aspects. Under each category, respondents were asked about their attitudes and preferences in hypothetical yet plausible scenarios. Section "The Ethical Questions and Findings" contains a detailed report of the findings for the ethical questions, followed by our commentary in section "Comments on Ethical Findings". Similarly, section "The Legal Questions and Findings" presents the findings for the legal questions, followed by our commentary in section "Comments on Legal Findings".

The Ethical Questions and Findings

1. Do you agree that for dealing with crisis cases (such as a brake failure situation) AVs should be designed and programmed always to ensure the safety of its driver and occupants only, but not other road users? (Agree / disagree)

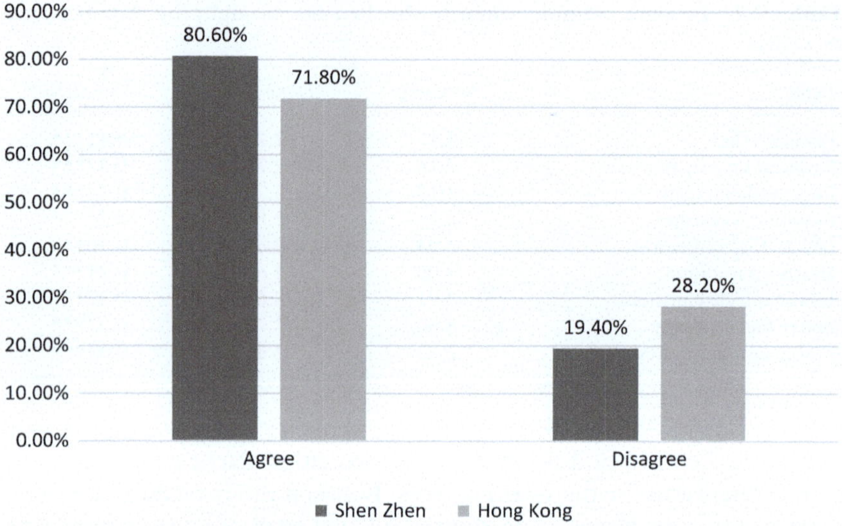

Fig. 9.1 AVs should be designed and programmed always to ensure the safety of its driver and occupants, but not other road users

Finding: When asked about their preference for the design and programming of AVs, 76.2% of the participants expressed agreement that scientists and engineers should prioritize the safety of drivers and passengers over other road users. Notably, a higher proportion of participants in Shenzhen (80.6%) held this viewpoint compared to those in Hong Kong (71.8%), and this difference was found to be statistically significant (chi2 = 10.68, $p < 0.001$) (see Fig. 9.1).

2. Would you like to purchase an AV which is designed and programmed to always spare the lives of the pedestrians by swerving onto a solid barrier in a break failure situation? (Yes / No)

Finding: In response to this question, 77.6% of the participants agreed to purchase an AV designed and programmed to always steer towards a solid obstacle to save pedestrians in the event of brake failure. The proportion was significantly higher in Shenzhen (82.4%) compared to Hong Kong (72.8%), and the difference was statistically significant (chi2 = 13.25, $p < 0.001$). (see Fig. 9.2).

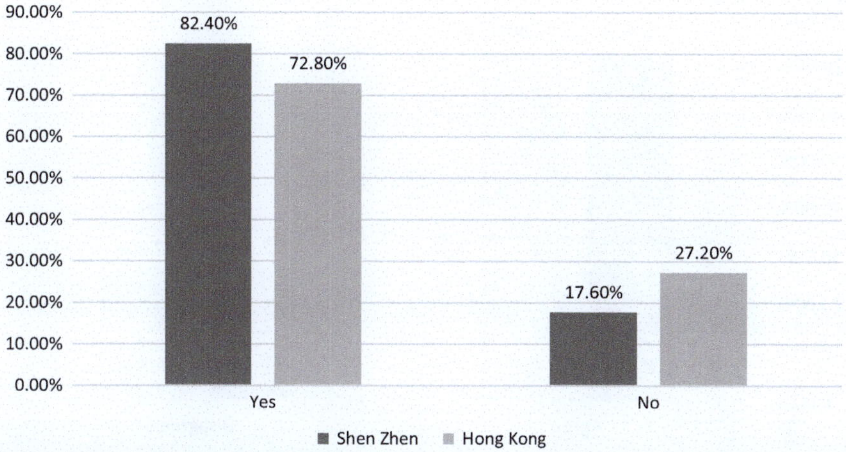

Fig. 9.2 Willingness to purchase an AV which is designed and programmed to always spare the lives of the pedestrians by swerving onto a solid barrier in a break failure situation

3. Do you agree that for dealing with a brake failure situation an AV should be designed and programmed to swerve onto a solid barrier which will kill or injure the driver and occupants but save multiple pedestrians? (Agree / disagree)

Finding: In response to this question, 72.7% of the participants agreed that AVs should be designed and programmed to steer towards a solid obstacle in the event of brake failure, which could potentially result in injury or death to the driver and passengers but save multiple pedestrians. The proportion was very close between Shenzhen (72.6%) and Hong Kong (72.8%), and the difference was statistically insignificant (chi2 = 0.01, $p > 0.05$). (see Fig. 9.3).

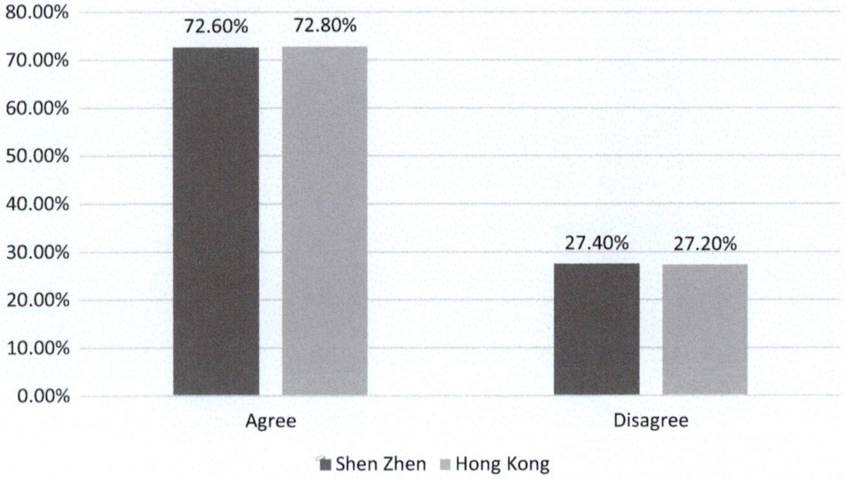

Fig. 9.3 AVs should be designed and programmed to steer towards a solid obstacle in the event of brake failure, which could potentially result in injury or death to the driver and passengers but save multiple pedestrians

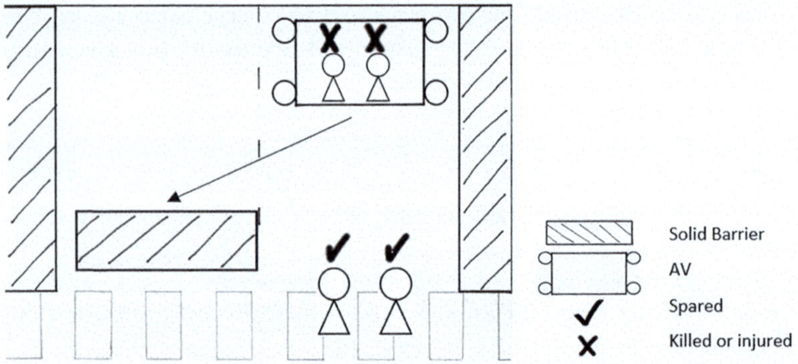

4. Do you agree that for dealing with a brake failure situation an AV should be designed and programmed to continue to run into multiple pedestrians, which will kill or injure the pedestrians but save the driver and occupants? (Agree / disagree)

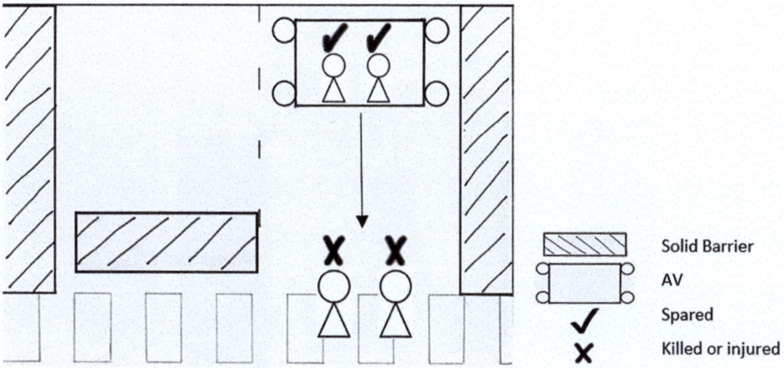

Finding: In response to this question, only 45.8% of the participants agreed that, in dealing with a brake failure situation, an AV should be designed and programmed to continue to run into multiple pedestrians, which will kill or injure the pedestrians but save the driver and occupants. Among the respondents in Shenzhen, 51.4% agreed, while in Hong Kong, only 40.2% agreed. (see Fig. 9.4).

5. If young people (such as aged 5–15) are the occupants in an AV, and same-number pedestrians on the road are healthy adults, do you agree that the AV should be designed and programmed to spare the lives of the occupants and never swerve onto a solid barrier to risk their lives, while the adult pedestrians may be killed or injured? (Agree /disagree)

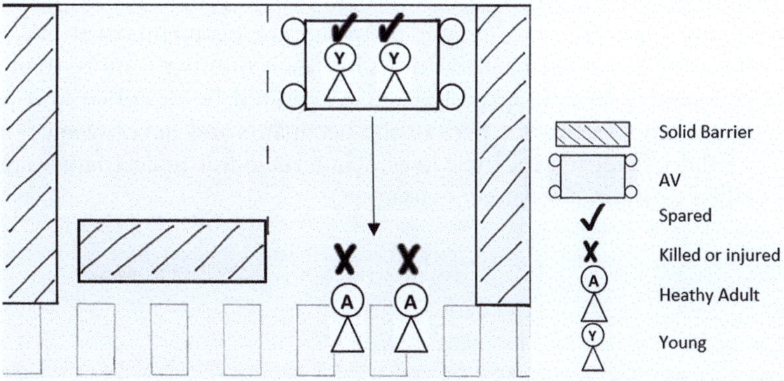

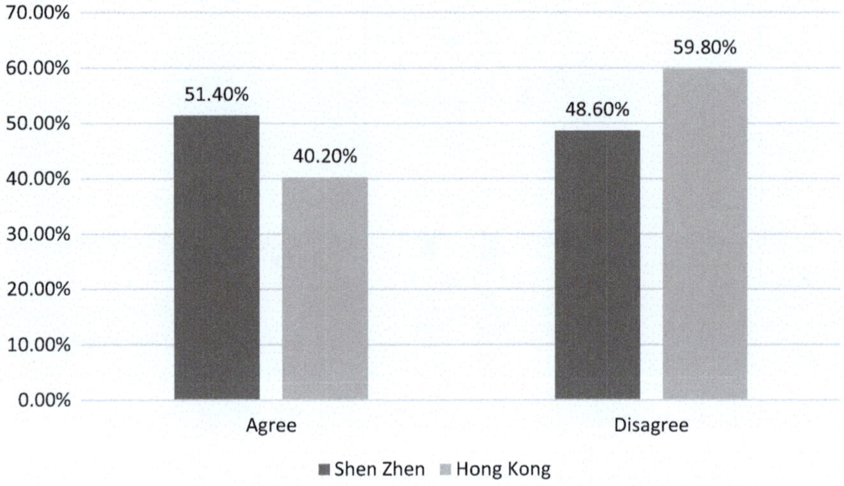

Fig. 9.4 In dealing with a brake failure situation, AVs should be designed and programmed to continue to run into multiple pedestrians, which will kill or injure the pedestrians but save the driver and occupants

Finding: In response to this question, 55.3% of the participants agreed that if young people are the occupants in an AV and same-number pedestrians on the road are healthy adults, the AV should be designed and programmed to spare the lives of the occupants. The proportion of agreement was significantly higher in Shenzhen (62%) compared to Hong Kong (48.6%), and the difference was statistically significant (chi2 = 18.16, $p < 0.001$). (see Fig. 9.5).

6. If young people (such as aged 5–15) are the occupants in an AV, and the pedestrians are healthy adults but their number is more than the occupants, do you agree that an AV should be designed and programmed to spare the lives of the occupants and never swerve onto a solid barrier to risk their lives, while the adult pedestrians may be killed or injured? (Agree / disagree)

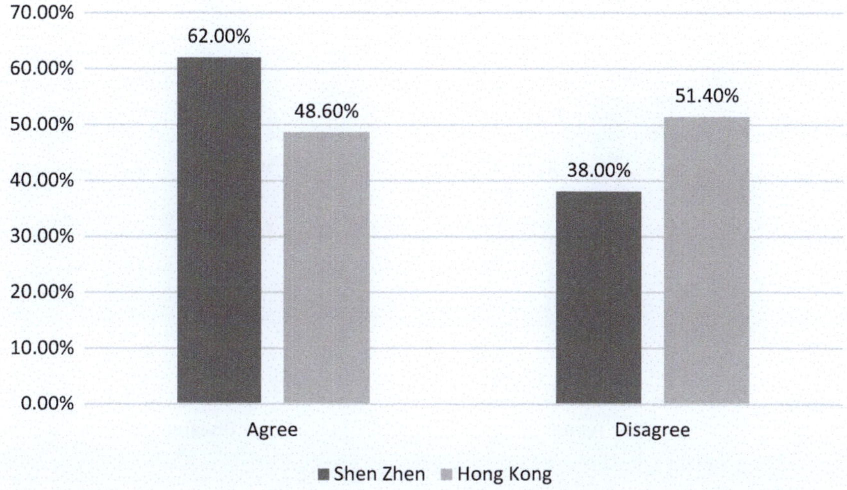

Fig. 9.5 If young people are the occupants in an AV and same-number pedestrians on the road are healthy adults, AVs should be designed and programmed to spare the lives of the occupants and never swerve onto a solid barrier to risk their lives, while the adult pedestrians may be killed or injured

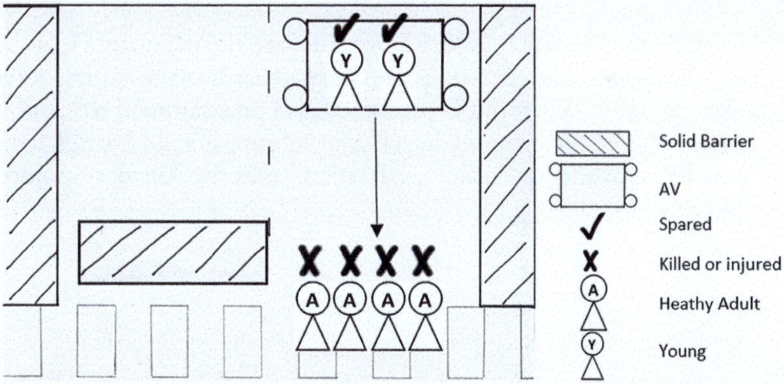

Finding: In response to this question, only 47.2% of the participants agreed that a self-driving car should be designed and programmed to never steer towards a solid obstacle to protect child passengers in the car, even if it meant endangering more healthy adult pedestrians on the road. Among the respondents in Shenzhen, 55.4% agreed, while in Hong Kong, only 39% agreed (chi2 = 26.98, $p < 0.001$). (See Fig. 9.6).

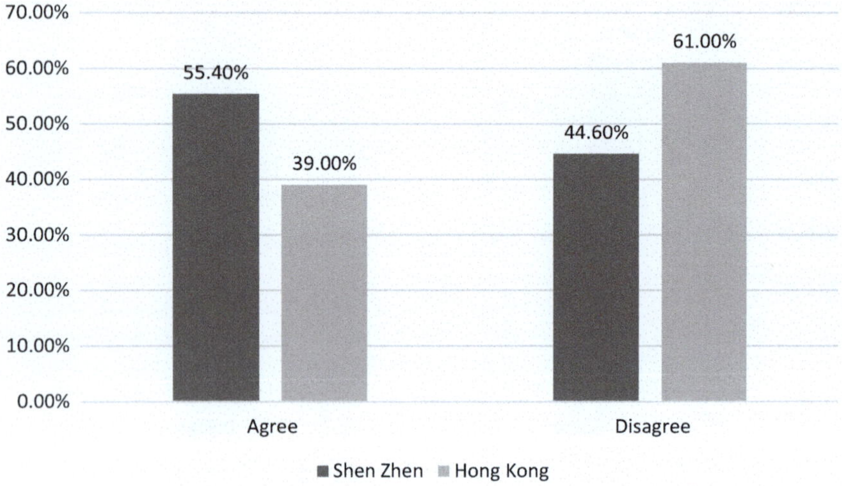

Fig. 9.6 If young people are the occupants in an AV and the pedestrians are healthy adults but their number is more than the occupants, AVs should be designed and programmed to spare the lives of the occupants and never swerve onto a solid barrier to risk their lives, while the adult pedestrians may be killed or injured

7. If elderly people (such as those over 65) are the occupants in an AV, and same-number pedestrians on the road are healthy adults, do you agree that the AV should be designed and programmed to spare the lives of the occupants and never swerve onto a solid barrier to risk their lives, while the adult pedestrians may be killed or injured? (Agree / disagree)

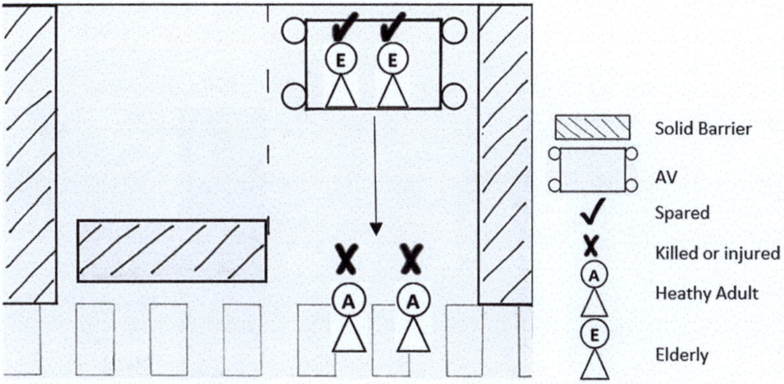

Finding: In response to this question, only 41.8% the participants agreed. Among the respondents in Shenzhen, 46.8% agreed, while in Hong Kong, only 36.8% agreed. The proportion of agreement was significantly higher in Shenzhen compared to Hong Kong (chi2= 10.28, $p < 0.001$). (See Fig. 9.7).

8. If elderly people (such as those over 65) are the occupants in an AV, and the pedestrians are healthy adults but their number is more than the occupants, do you agree that the AV should be designed and programmed to spare the lives of the occupants and never swerve onto a solid barrier to risk their lives, while the adult pedestrians may be killed or injured? (Agree / disagree)

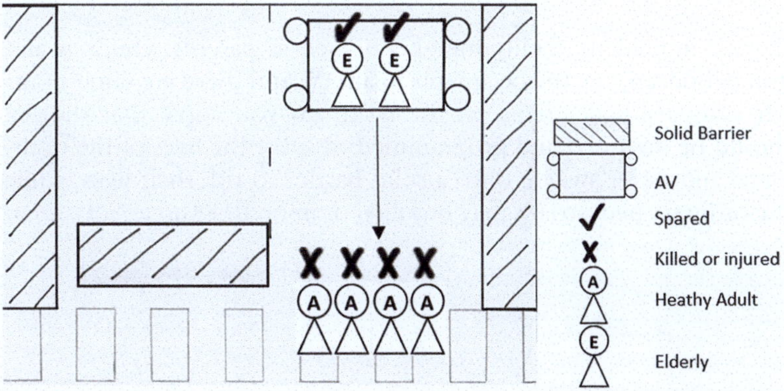

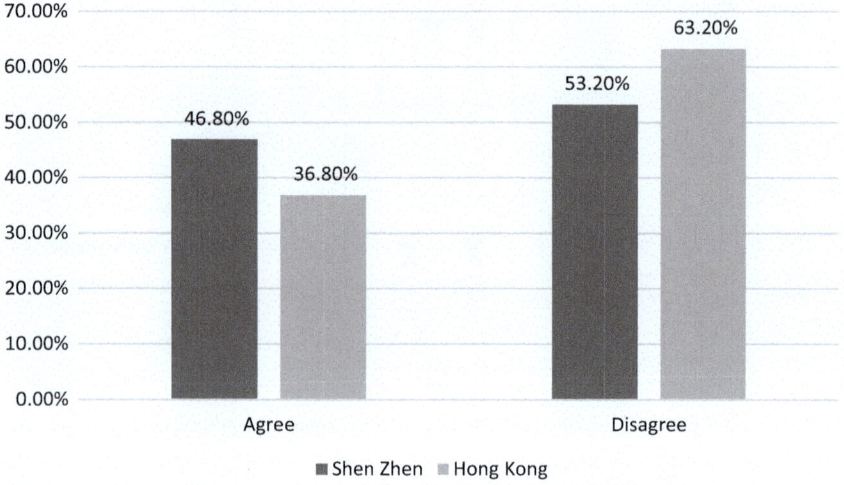

Fig. 9.7 If elderly people are the occupants in an AV and same-number pedestrians on the road are healthy adults, AVs should be designed and programmed to spare the lives of the occupants and never swerve onto a solid barrier to risk their lives, while the adult pedestrians may be killed or injured

Finding: In response to this question, only 37.5% of the participants agreed to spare the lives of the occupants if more heathy adults pedestrian would be killed or injured. Among the respondents in Shenzhen, 42.8% agreed, while in Hong Kong, only 32.2% agreed. The difference was statistically significant, with a higher proportion of agreement in Shenzhen compared to Hong Kong (chi2 = 11.99, $p < 0.001$). (See Fig. 9.8).

9. If your immediate family members (such as parents, children and grandchildren) are the occupants in an AV, and there are same-number ordinary pedestrians on the road, do you agree that the AV should be designed and programmed to spare the lives of the occupants and never swerve onto a solid barrier to risk their lives, while the ordinary pedestrians may be killed or injured? (Agree / disagree)

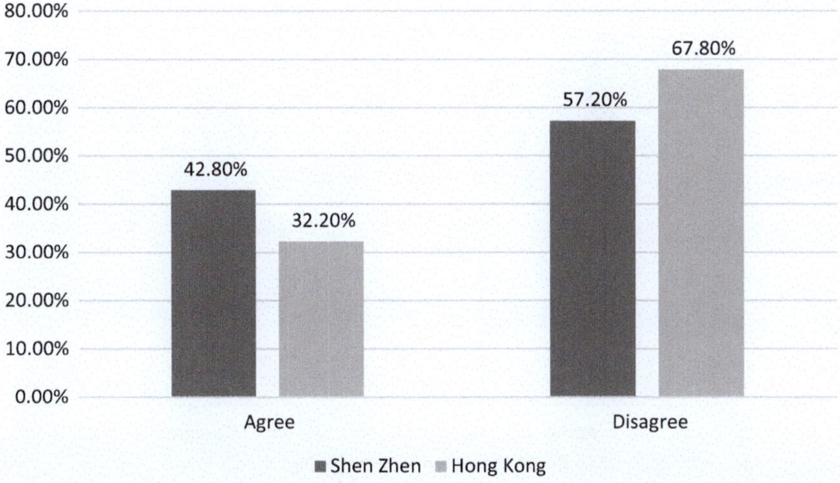

Fig. 9.8 If elderly people are the occupants in an AV and the pedestrians are healthy adults but their number is more than the occupants, AVs should be designed and programmed to spare the lives of the occupants and never swerve onto a solid barrier to risk their lives, while the adult pedestrians may be killed or injured

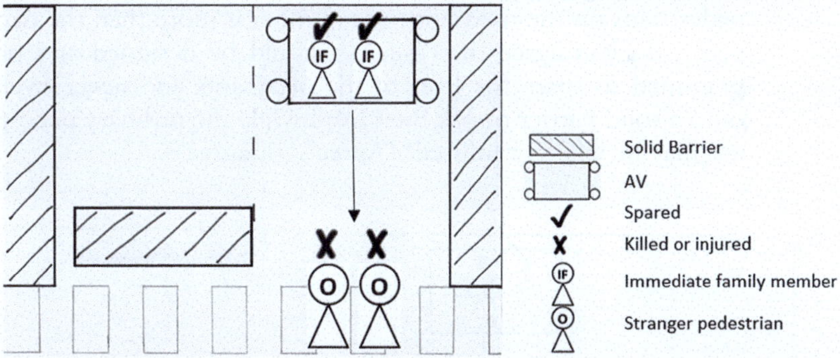

Finding: In response to this question, 59.8% participants agreed to protect the family member in the car, even if it meant endangering a same number of pedestrians on the road. The proportion of agreement was significantly higher in Shenzhen (66%) compared to Hong Kong (53.6%), and the difference was statistically significant (chi2 = 15.99, $p < 0.001$). (See Fig. 9.9).

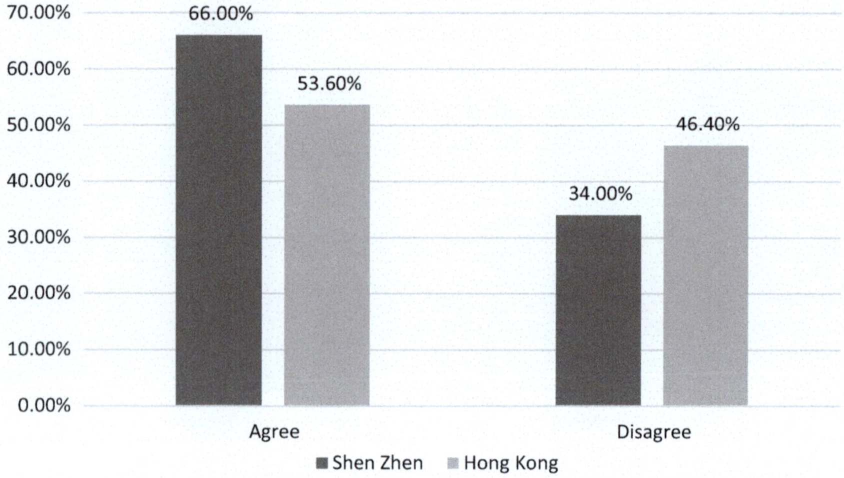

Fig. 9.9 If your immediate family members are the occupants in an AV and there are same-number ordinary pedestrians on the road, AVs should be designed and programmed to spare the lives of the occupants

10. If your immediate family members (such as parents, children and grandchildren) are the occupants in an AV, and there are ordinary pedestrians on the road but their number is more than the occupants, do you agree that the AV should be designed and programmed to spare the lives of the occupants and never swerve onto a solid barrier to risk their lives, while the ordinary pedestrians may be killed or injured? (Agree / disagree)

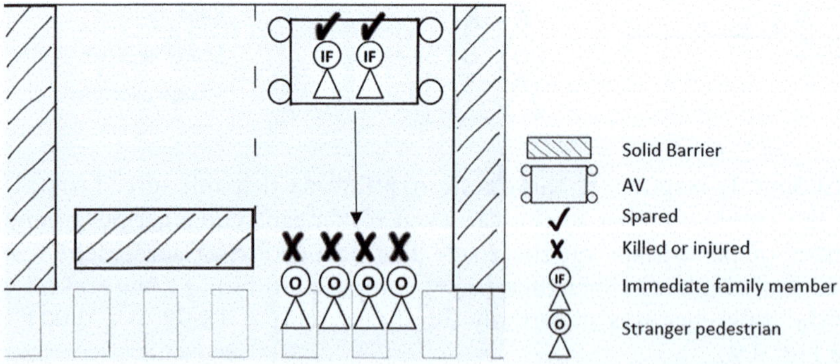

Finding: In response to this question, 52.9% of the participants agreed to protect the family member in the car. Among the respondents in Shenzhen, 58% agreed, while in Hong Kong, only 47.8% agreed. The difference was statistically significant, with a higher proportion of agreement in Shenzhen compared to Hong Kong (chi2 = 10.44, $p < 0.01$). (See Fig. 9.10)

11. In a brake failure situation in which an AV must either run into 3 pedestrians or swerve onto 2 other pedestrians, do you agree that the AV should be designed and programmed to swerve onto the 2 pedestrians in order to spare the 3 pedestrians? (Agree / disagree)

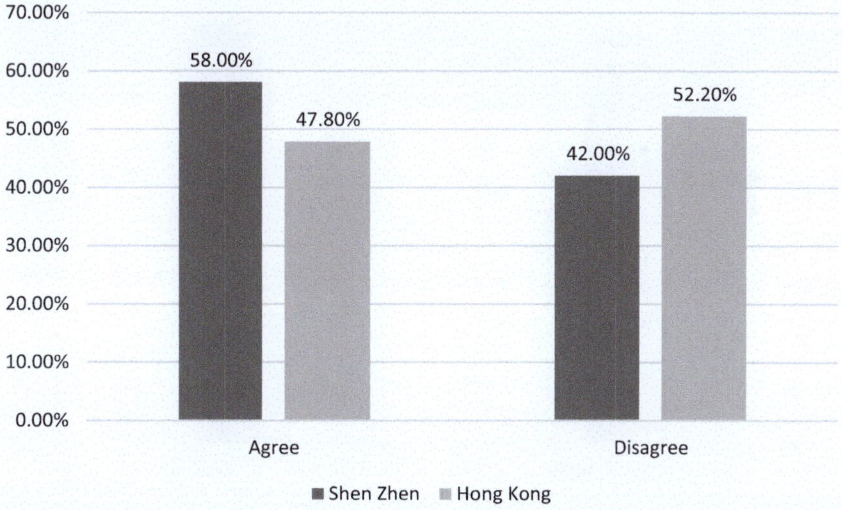

Fig. 9.10 If your immediate family members are the occupants in an AV and there are ordinary pedestrians on the road but their number is more than the occupants, AVs should be designed and programmed to spare the lives of the occupants

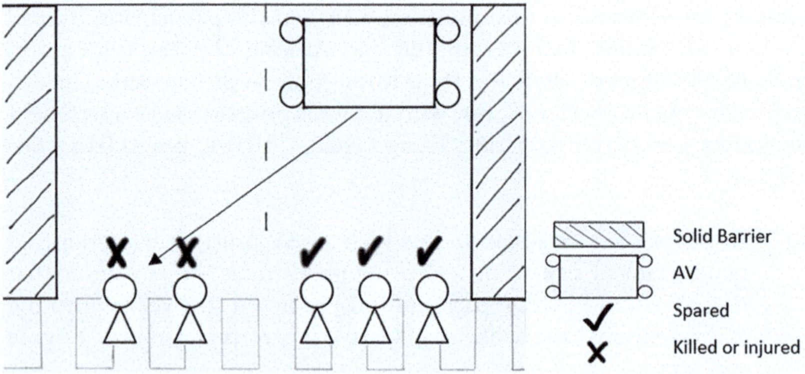

Finding: In response to this question, 73.3% of the participants agreed that AVs should be designed and programmed to steer towards two pedestrians to save three pedestrians in the event of brake failure. There was no significant difference between the proportion of agreement in Shenzhen (74.6%) and Hong Kong (72%). (See Fig. 9.11).

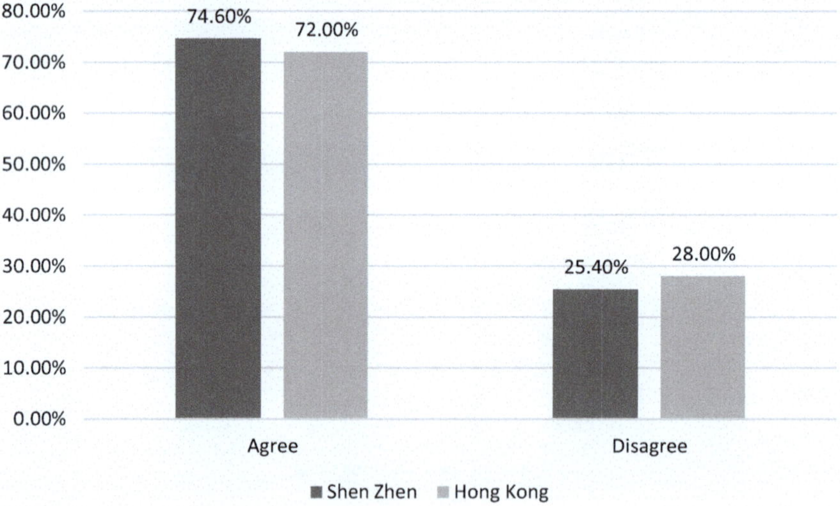

Fig. 9.11 In a brake failure situation in which an AV must either run into 3 pedestrians or swerve onto 2 other pedestrians, AVs should be designed and programmed to swerve onto the 2 pedestrians in order to spare the 3 pedestrians

12. In a brake failure situation in which an AV must run into one of the following pedestrians, which pedestrian do you think an AV should be designed and programmed to swerve onto in order to save other pedestrians? Please rank (1, 2, 3, 4) (1=must be swerved onto; 4 = must not be swerved onto)

    (a) children
    (b) elderly individuals
    (c) adult males
    (d) adult females

Finding: In response to this question, the participants ranked adult males as the group to be hit first, followed by the elderly individuals, adult females, and children. There was no significant difference in the ranking order between Shenzhen and Hong Kong. (See Fig. 9.12).

13. In a brake failure situation in which an AV must run into either one of the following groups of pedestrians, please choose which group you think the AV should be designed and programmed to swerve onto in order to save the remaining group of pedestrians.

    (a) pedestrians who follow the road traffic rules;

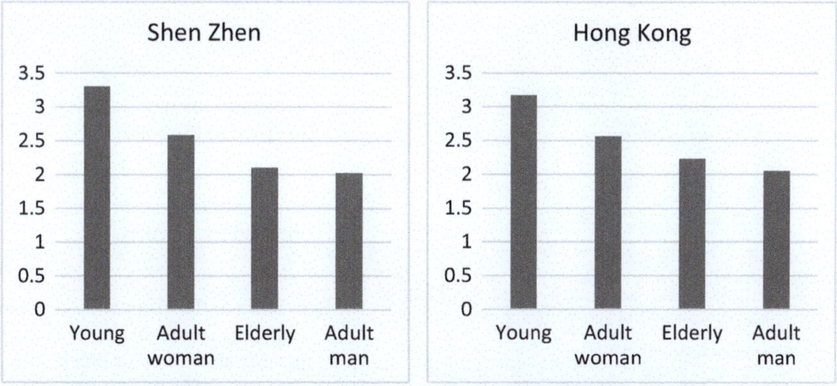

Fig. 9.12 In a brake failure situation in which an AV must run into one of the following pedestrians, which pedestrian an AV should be designed and programmed to swerve onto in order to save other pedestrians?

(b) pedestrians who break the road traffic rules;
(c) Keep to the original route, regardless of the pedestrians follow or break the road traffic rules

Finding: In response to this question, 64.5% of the participants chose pedestrians who violate traffic rules, 25.4% chose to continue driving straight, and the remaining 10.1% chose pedestrians who follow traffic rules. There was no significant difference in the proportion of choices between Shenzhen and Hong Kong (Pearson chi-2 = 4.76, $p > 0.05$). (See Fig. 9.13).

14. Do you agree that AVs used for public transport purposes (namely, having a large group of occupants, such as public buses, school buses, and elderly home buses) should be designed and programmed to always protect the occupants rather than the pedestrians in a crisis situation such as a brake failure case (e.g., it should never be swerved onto a solid barrier to risk the lives of the occupants)? (Agree / disagree)

Finding: In response to this question, 67.5% of the participants agreed to protect the occupants in the AVs utilized for public transport purposes. The proportion of agreement was significantly higher in Shenzhen (73.8%)

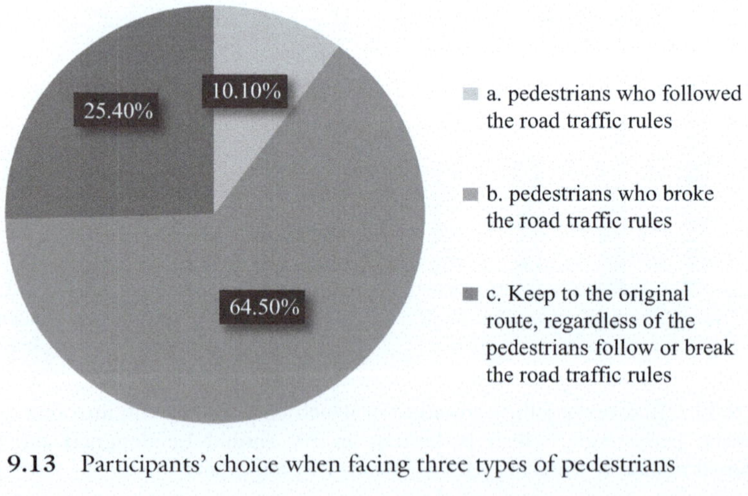

Fig. 9.13 Participants' choice when facing three types of pedestrians

compared to Hong Kong (61.2%), and the difference was statistically significant (chi2= 18.09, p < 0.001). (See Fig. 9.14).

15. Do you agree that an AV for individual- or family- use (namely, having only one occupant or a small group of occupants) should be designed and programmed always to swerve onto a solid barrier rather than to select human targets in a crisis in order to save the lives of multiple pedestrians? (Agree / disagree)

Finding: In response to this question, 75.5% of the participants agreed. There was no significant difference in the proportion of agreement between Shenzhen (76.4%) and Hong Kong (74.6%) (chi2 = 0.44, p > 0.05). (See Fig. 9.15).

16. If AVs for individual- or family- use have been designed and programmed always to swerve onto a solid barrier rather than to select human targets in a crisis in order to save the lives of multiple pedestrians, do you agree that such an AV should always keep a steering wheel available for a human driver to take over and control at any time he/she wishes (including at the time of brake

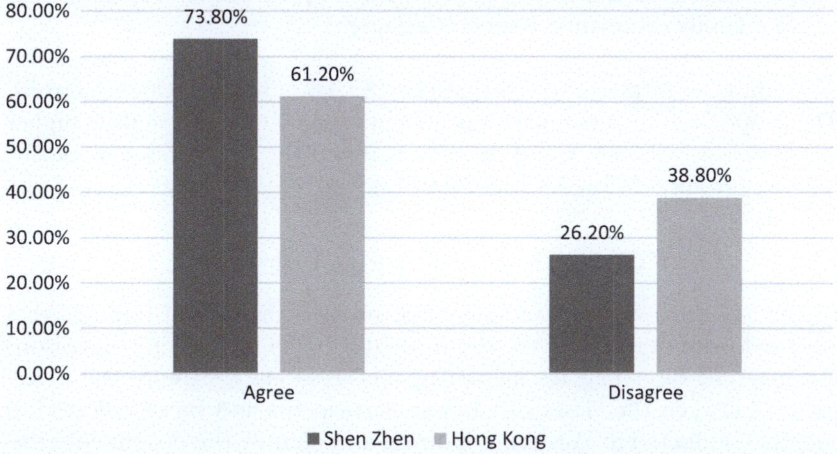

Fig. 9.14 AVs used for public transport purposes should be designed and programmed to always protect the occupants rather than the pedestrians in a crisis situation

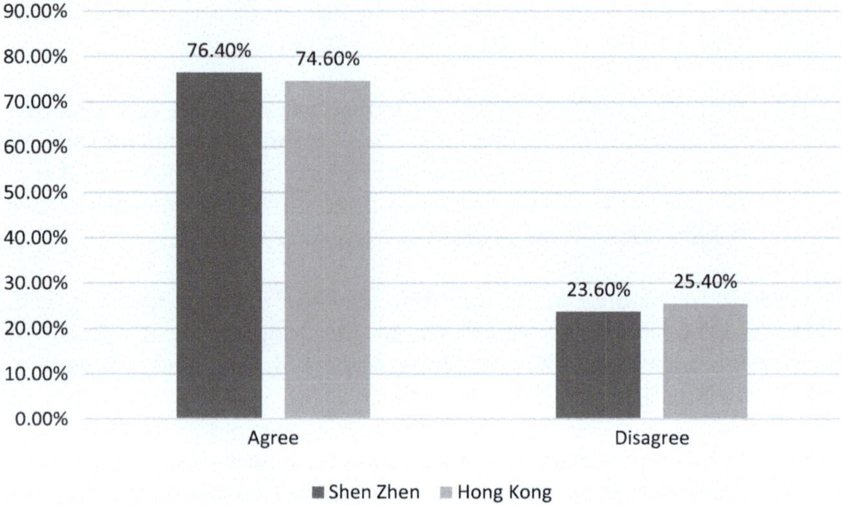

Fig. 9.15 An AV for individual- or family- use should be designed and programmed always to swerve onto a solid barrier rather than to select human targets in a crisis in order to save the lives of multiple pedestrians

> failure) so that she/he may decide to protect the lives of his/her family occupants? (Agree / disagree)

Finding: In response to this question, 87.1% of the participants agreed. The proportion of agreement was 85% in Hong Kong and slightly higher in Shenzhen (89.2%). The difference between the two cities was statistically significant (chi2 = 3.92, $p < 0.05$). (See Fig. 9.16).

Comments on Ethical Findings

In this section, we will examine three noteworthy ethical findings that emerged from our survey. We aim to shed light on the underlying reasons for apparent discrepancies in participants' responses, explore the divergences between the views of Chinese participants and those reflected in surveys conducted in Western countries, and identify the prominent ethical values embodied in the choices made by Chinese participants.

There appears to be a discrepancy between the responses to questions (1) and (2). In question 1), 76.2% of the participants supported the idea

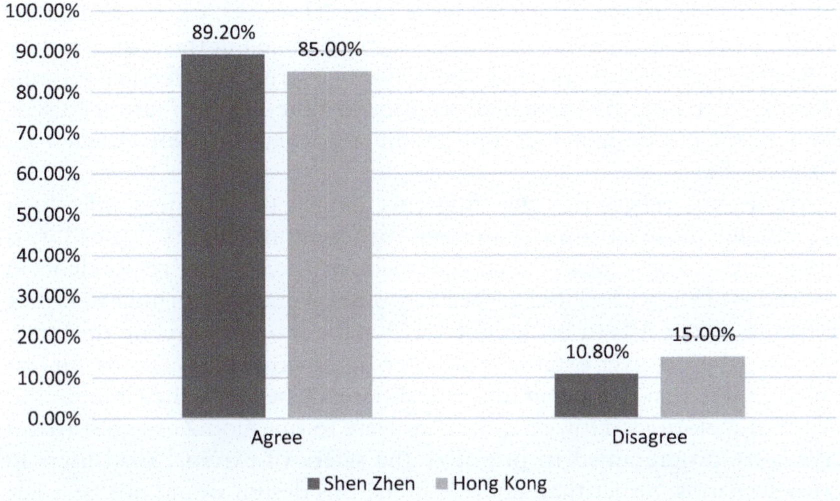

Fig. 9.16 An AV should always keep a steering wheel available for a human driver to take over and control at any time he/she wishes (including at the time of brake failure) so that she/he may decide to protect the lives of his/her family occupants

that scientists and engineers should prioritize the safety of AVs' drivers and passengers over other road users in potential crisis situations, such as brake failure. However, when it comes to the willingness to purchase an AV that is designed and programmed to prioritize saving pedestrians by swerving onto a solid barrier in a brake failure scenario (question 2)), the responses are surprisingly different. In fact, 77.6% of the participants expressed their readiness to purchase such an AV. This contradiction raises intriguing questions about participants' perspectives. Despite the statistically significant differences between Hong Kong and Shenzhen, we will set aside these differences for now, as the majority views in both cities align.

It is indeed valuable to examine the differences between the views of Chinese participants and those observed in surveys conducted in Western countries. For instance, surveys conducted in certain Western countries have revealed that a majority of individuals believe it is morally correct to pre-program AVs to minimize casualties in emergency situations. This means that if the number of internal occupants is fewer than the number of external road users, AVs should be programmed to swerve onto a barrier in order to spare the lives of the latter. However, when asked if they would personally purchase or rent AVs programmed in this manner, their willingness diminishes. Instead, the majority express a preference for purchasing or renting AVs designed to prioritize the safety of internal users, even if such a design does not minimize casualties (Bennefon et al., 2016, 1574).

We do not believe that this difference implies that Chinese individuals are morally more altruistic than their Western counterparts. Instead, two other factors may explain the choices made by the Chinese respondents in questions (1) and (2). First, as their response to question 1) indicates, they generally prefer AVs to be designed with a focus on prioritizing the safety of internal users over external road users in emergency situations. In this aspect, there may not be an inherent difference between the Chinese and Western majorities. However, their answers to question 2) suggest that if AVs were programmed to prioritize the safety of external road users in crisis scenarios, rather than internal users, this design would not be a significant deterrent for them to utilize such AVs. One plausible reason for this could be their perception that emergency situations are infrequent occurrences. Consequently, they believe that even if AVs were programmed to prioritize external road users, they would still be as safe as or potentially safer than when using human-driven vehicles.

Moreover, the ethos of Confucian meritocracy may play a significant role in shaping these preferences.[1] In Confucian culture, there is a belief that while everyone is born with equal potential for virtue, not everyone possesses actualized virtues to the same degree. Individuals who have demonstrated excellence in intellectual pursuits and cultivated their moral virtues are seen as setting positive examples for others in society to emulate. Therefore, if individuals with notable merits determine that AVs should be programmed to prioritize the protection of external users in emergency situations, their decisions are likely to be trusted and followed by others. While it cannot be assumed that AV scientists and engineers are inherently virtuous individuals solely responsible for determining the design of AVs, they are expected to collaborate with other relevant experts in society to carry out their work diligently. Instead of relying solely on individualistic and independent judgment when deciding whether to

[1] For a useful introduction to Confucian meritocracy for contemporary society, see Bell and Li (eds.), Bell & Li, 2013.

purchase or rent AVs, Chinese people tend to place their trust in these knowledgeable experts within the framework of Confucian meritocracy.

It is important to acknowledge that certain cultural differences exist between Hong Kong and Shenzhen, despite both cities being influenced by Confucian values. A notable contrast can be observed in the responses to questions (9) and (10). Question (9) inquires whether participants agree that an AV should be designed to prioritize the protection of the occupants, who are their immediate family members (such as parents, children, and grandchildren), and this means not swerving onto a solid barrier to endanger their lives, even if an equal number of outside pedestrians, who are strangers, would be injured or killed. Nearly 60% of the participants responded positively to this question. Although more participants from Shenzhen chose this option compared to Hong Kong, the figure in Hong Kong still exceeded 50%. Hence, the majority of people in both mainland China and Hong Kong express a clear preference for prioritizing the safety of their family members over other individuals when the numbers are equal.

However, when considering question 10, which presents a similar scenario to question 9 but with a higher number of outside strangers compared to the occupant family members, a significant difference emerges between the choices of respondents from Shenzhen and Hong Kong. While 58% of the respondents in Shenzhen gave a positive answer (indicating that the AV should continue to prioritize the safety of the occupants and not swerve onto a solid barrier to endanger their family members), only 47.8% of the respondents in Hong Kong shared the same view. In other words, more respondents from Hong Kong concluded that, in this situation, the AV should be programmed to swerve onto a solid barrier in order to protect pedestrians whose number exceeds that of the occupant family members.

These findings may indicate that the people of Hong Kong have been influenced by both Confucian familism and British utilitarianism, while the impact of British utilitarianism in mainland China is less pronounced. Utilitarianism emphasizes the moral importance of maximizing overall pleasure in decision-making. If more people are injured, it would definitely decrease total pleasures and increase total pains, regardless of whether they are family members.[2] On the other hand, Confucianism

[2] For the utilitarian view of emphasizing the importance of impartiality and treating all individuals equally in order to maximize overall happiness, see a broadly influential paper by Singer (1972)).

emphasizes the significant moral obligation one has to care for their family members compared to others. This is not because Confucianism asserts that family members are inherently more valuable than others, but rather because these relationships hold moral significance and should be considered in ethical decision-making. According to the Confucian perspective, if individuals do not give preferential treatment to their immediate family members, the world would become morally impoverished, and authentic human connections would be lost.[3]

The survey responses to question (15) indicate that 75.5% of the participants, without any significant difference between Hong Kong and Shenzhen, agree that a small AV designed for individual or family use should be programmed to swerve onto a solid barrier rather than select human targets in a crisis, prioritizing the lives of multiple pedestrians. This survey result shows a consensus among the majority of the participants regarding the moral choice of sacrificing oneself and even some family members to spare the lives of others. This sentiment is consistent between the two regions surveyed. Regarding question (16), 87.1% of the participants (89.2% in Shenzhen and 85% in Hong Kong) support the inclusion of a steering wheel in such AVs. This provision would allow a human driver to take control and make decisions at any time, including during brake failure or other critical situations. The participants' overwhelming agreement highlights their belief that maintaining a steering wheel empowers them to protect the lives of their family members in the AV according to their own judgment.

The response to question 15 indicates that the majority of Chinese participants view sacrificing oneself to spare others' lives as a virtuous act, particularly considering that such AVs are typically used by individual owners. They believe it is appropriate for AVs to be programmed to prioritize the protection of external road users over the vehicle occupants in emergency situations. Simultaneously, they strongly support the inclusion of a steering wheel in AVs, enabling them to assume control and safeguard the lives of their family members according to their own judgment during crisis scenarios. This prevailing preference in both Hong Kong and Shenzhen not only highlights the participants' recognition of their moral agency and the lack thereof in AVs, but also underscores their commitment to a family-oriented ethics. They firmly believe in their moral

[3] For a general introduction to Confucian ethics as well as the doctrine of care with distinctions, see Wong (2023)).

obligation to care for and protect their family members, even if it contradicts the utilitarian calculation of maximizing overall happiness in the design of AVs for handling exceptional circumstances. This collective stance reflects the importance people place on their ability to be in control of AVs when they deem it necessary, aligning with their values as moral agents and their dedication to family-centric values.

The Legal Questions and Findings

Question 1: How concerned are you about the following safety issues related to highly automated vehicles (Level 4 and Level 5)?

Finding (Fig. 9.17): In an investigation into regional differences concerning safety issues associated with highly autonomous vehicles (Levels 4 and 5), we found statistically significant differences between respondents from Shenzhen and Hong Kong on several key aspects. The scores were measured on a scale from 1 to 4, where 1 represents "not at all concerned," 2 represents "slightly concerned," 3 represents "moderately concerned," and 4 represents "very concerned." "Safety consequences of equipment or system failure" showed the highest concern in Shenzhen (M = 3.25) compared to Hong Kong (M = 3.00), with a mean difference of

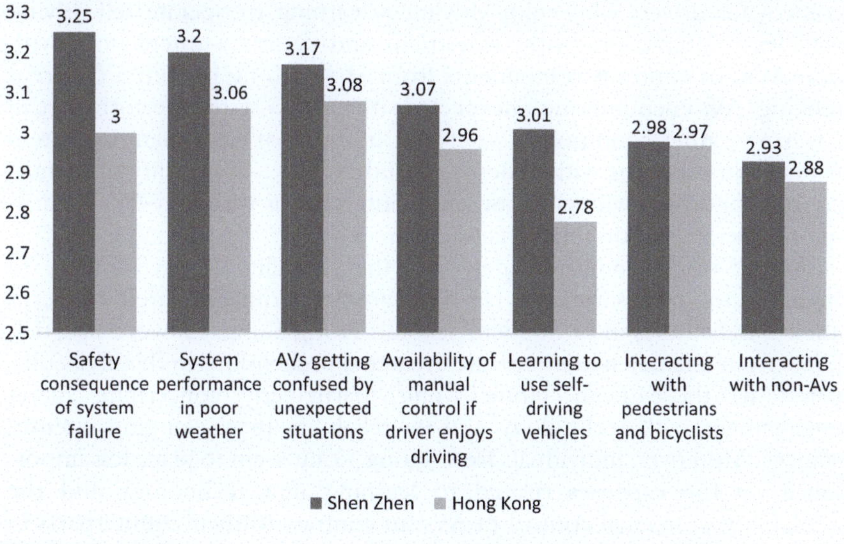

Fig. 9.17 Safety issues concerns related to highly automated vehicles

0.25 ($p < 0.001$). This score of 3.25 indicates that respondents in Shenzhen expressed a "moderate to high" level of concern regarding this issue. Similar significant concerns were noted for "learning to use self-driving vehicles", with Shenzhen reporting higher apprehension ($M = 3.01$) than Hong Kong ($M = 2.78$) with a mean difference of 0.23 ($p < 0.001$). "System performance in poor weather" (mean difference = 0.14, $p < 0.01$) and "availability of manual control if the driver enjoys driving" (mean difference = 0.11, $p < 0.05$) were also areas where Shenzhen showed significantly greater concern.

On the other hand, some issues did not exhibit significant differences between the two cities. These included "AVs getting confused by unexpected situations" (mean difference = 0.09, $p > 0.05$), "interacting with non-AVs" (mean difference = 0.06, $p > 0.05$), and "interacting with pedestrians and bicyclists" (mean difference = 0.01, $p > 0.05$), indicating similar levels of concern in both locations with regard to these aspects.

Moreover, a distinction between participants in Shenzhen and Hong Kong is also evident in terms of the level of their concerns. According to the responses from Shenzhen, their concerns regarding autonomous vehicles, ranked from highest to lowest, are as follows: safety implications of equipment or system failure > system performance in adverse weather conditions > AVs encountering unexpected situations > availability of manual control for drivers who enjoy driving > learning to operate self-driving vehicles > interactions with pedestrians and cyclists > interactions with non-AVs. In contrast, respondents from Hong Kong present a different ranking: AVs encountering unexpected situations > system performance in adverse weather conditions > safety implications of equipment or system failure > interactions with pedestrians and cyclists > availability of manual control for drivers who enjoy driving > interactions with non-AVs > learning to operate self-driving vehicles.

As the results indicate, safety and the reliability of performance of highly autonomous vehicles (H-AVs) are the primary shared concerns among individuals in both Shenzhen and Hong Kong. However, participants from Shenzhen demonstrate greater apprehensions in this area compared to those from Hong Kong, particularly concerning direct engagement with technology and its reliability in challenging circumstances. Moreover, individuals from Hong Kong seem to place less importance on the expenses related to learning new technology and the availability of manual control of AVs, in contrast to their counterparts in Shenzhen.

These findings may suggest wider cultural or infrastructural differences that could impact the acceptance and regulation of autonomous vehicle technology. From a practical perspective, these results suggest that policy-makers and manufacturers should consider regional differences in safety perceptions. They may need to strengthen educational campaigns and regulatory frameworks in areas where there is more scepticism towards autonomous vehicles.

Question 2: How concerned are you about the following data privacy breach threats related to AVs?

Finding (Fig. 9.18): In response to data privacy concerns, our findings reveal statistically significant differences across all surveyed dimensions, suggesting varying levels of apprehension between the two cities. The scores were measured on a scale from 1 to 4, where 1 represents "not at all concerned," 2 represents "slightly concerned," 3 represents "moderately concerned," and 4 represents "very concerned." Regarding "personal information privacy concerning geographic tracking", respondents from Shenzhen exhibited a significantly higher level of concern ($M = 3.25$) compared to those from Hong Kong ($M = 3.02$), with a mean difference of 0.23 ($p < 0.001$). Similarly, "vehicle security" was another critical area where Shenzhen respondents reported greater apprehension ($M = 3.42$) than their Hong Kong counterparts ($M = 3.23$), yielding a mean

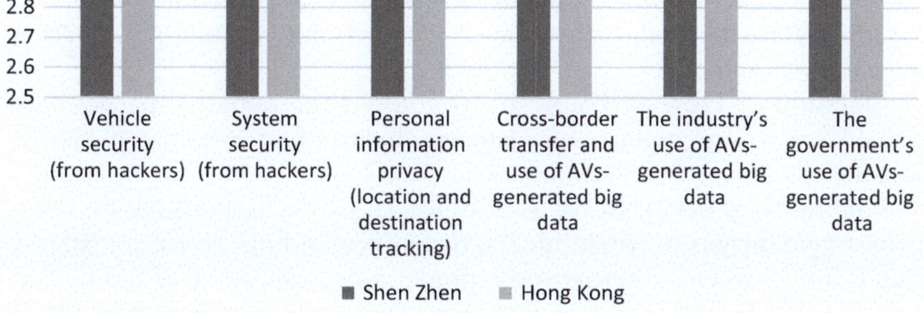

Fig. 9.18 Data privacy concerns related to AVs

difference of 0.18 ($p < 0.01$). "System security" also showed a noticeable difference, albeit smaller, with Shenzhen at a mean of 3.40 compared to Hong Kong's 3.30, resulting in a mean difference of 0.10 ($p < 0.05$).

Moreover, "the industry's utilization of AVs-generated big data" was perceived with greater concern in Shenzhen ($M = 3.07$) compared to Hong Kong ($M = 2.94$), indicating a difference of 0.13 ($p < 0.01$). Similarly, worries regarding "the government's utilization of AVs-generated big data" also showed higher apprehension in Shenzhen ($M = 3.07$) than in Hong Kong ($M = 2.90$), with a mean difference of 0.16 ($p < 0.01$). The most significant discrepancy was observed in "the cross-border transfer and utilization of AVs-generated big data", where Shenzhen recorded a mean of 3.09 in contrast to Hong Kong's 2.85, demonstrating a mean difference of 0.24 ($p < 0.001$).

The notable variations across all dimensions highlight a consistently higher level of concern in Shenzhen regarding various aspects of data privacy within the realm of autonomous vehicle technology. Specifically, threats from hackers to vehicle and system security, as well as the tracking of AV users' location and destination information, rank among the primary data privacy and security worries. The remaining three aspects (cross-border transfer and utilization of big data generated by autonomous vehicles, industry utilization of such data, and government utilization of such data) are of relatively low concern in both Shenzhen and Hong Kong.

This trend may indicate broader disparities in privacy expectations and data protection regulations between the two cities. The increased sensitivity in Shenzhen towards local and cross-border data privacy issues could impact the strategies for implementing AV technologies and necessitate tailored educational and regulatory measures to effectively address these concerns. These findings emphasize the significance of taking regional variations in data privacy apprehensions into account when devising plans for the deployment and governance of emerging technologies such as autonomous vehicles.

Question 3: How agreeable are you about the potential parties who should pay compensation for damages incurred in automated vehicle accidents?

Finding (Fig. 9.19): In terms of potential parties responsible for compensating damages in AV accidents, the survey findings reveal statistically significant differences across most dimensions, indicating differing perspectives on liability for compensation between Shenzhen and Hong

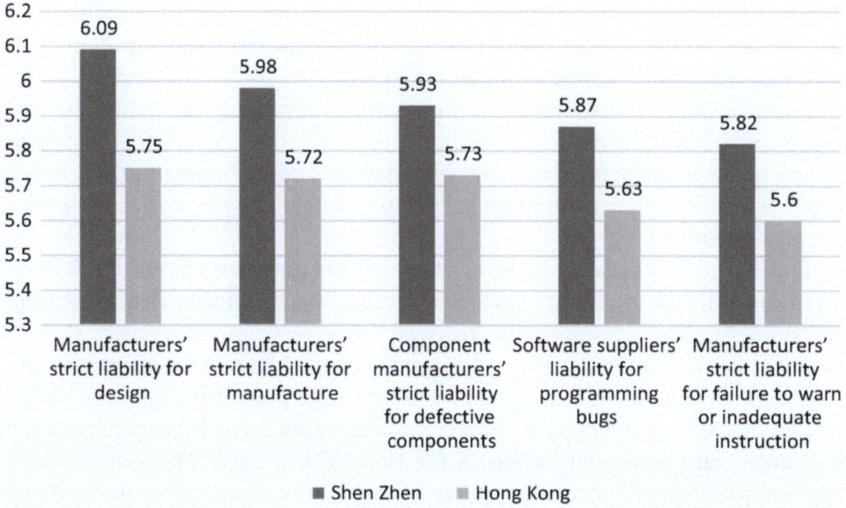

Fig. 9.19 The top five parties that participants believe should pay compensation for damages

Kong. The scores were measured on a scale from 1 to 7, where 1 represents "total disagreed," 2 represents "very disagreed," 3 represents "slightly disagreed," 4 represents "neutral," 5 represents "slightly agreed," 6 represents "very agreed," and 7 represents "total agreed." Particularly, significant disparities were noted in critical areas regarding manufacturers' responsibilities, such as "strict liability for defect in manufacture of vehicles" (mean difference = 0.26, $p < 0.001$), "failure to warn or inadequate instruction" (mean difference = 0.22, $p < 0.01$), "strict liability for defective components" (mean difference = 0.20, $p < 0.01$), and "strict liability for defective design" (mean difference = 0.34, $p < 0.001$). These findings suggest a stronger consensus in Shenzhen regarding the accountability of manufacturers in various aspects of AV production and design failures.

Furthermore, regarding "users' liability for the misuse", a slight yet statistically significant difference was observed, with Shenzhen showing higher agreement (M = 5.68) compared to Hong Kong (M = 5.53), indicating a mean difference of 0.15 ($p < 0.05$). Moreover, "the operators' strict liability for incorrect GPS and transportation information" was perceived significantly higher in Shenzhen (M = 5.78) than in Hong Kong (M = 5.57), demonstrating a mean difference of 0.21 ($p < 0.01$). "The government's no-fault regime for limited compensation" also garnered higher agreement in Shenzhen ($M = 5.24$) compared to Hong Kong ($M = 5.04$), with a difference of 0.20 ($p < 0.01$).

Nevertheless, there were no significant differences in "the owners' vicarious liability for misuse" (mean difference = 0.01, $p > 0.05$) and "the sellers' joint liability for defective automated vehicles" (mean difference = 0.01, $p > 0.05$), suggesting similar perspectives on these particular liabilities in both cities.

Despite differences in the level of consensus between respondents from Shenzhen and Hong Kong, they hold a common belief that manufacturers of AVs and their components, software suppliers, and operators should bear responsibility for compensating victims injured in AV accidents resulting from defects in AV design, manufacturing, programming, warnings, instructions, and operations. They coincidentally have high expectations and anticipate potential liabilities for the AV industry. The government appears to serve as a secondary party for providing compensation for damages arising from AV accidents. While acknowledging that AV users should take responsibility for accidents caused by misuse of vehicles, they perceive other potentially liable parties, such as sellers and AV owners, as less deserving of blame.

Question 4: How agreeable are you to the following reasons for liability exemption or reduction related to AVs?

Finding (Fig. 9.20): In response to the question, participants from both Hong Kong and Shenzhen demonstrate a high level of agreement that state-of-the-art immunity and users' contributory negligence in the misuse of AVs are grounds for exemption or reduction of liability. The scores were measured on a scale from 1 to 7, where 1 represents "total disagreed," 2 represents "very disagreed," 3 represents "slightly disagreed," 4 represents "neutral," 5 represents "slightly agreed," 6 represents "very agreed," and 7 represents "total agreed." Participants from Shenzhen exhibited a higher level of agreement ($M = 4.66$) compared to those from Hong Kong ($M = 4.12$). This suggests that participants from Shenzhen are more inclined to accept that current technological and scientific limitations at the time of an AV's release could justify exemptions from liability. Similarly, the concept of user contributory negligence in the misuse of AVs garnered greater agreement in Shenzhen ($M = 5.25$) than in Hong Kong ($M = 4.57$). This indicates that respondents from Shenzhen are more accepting of reducing or exempting the defendant's liability if the user's actions contribute to the incident.

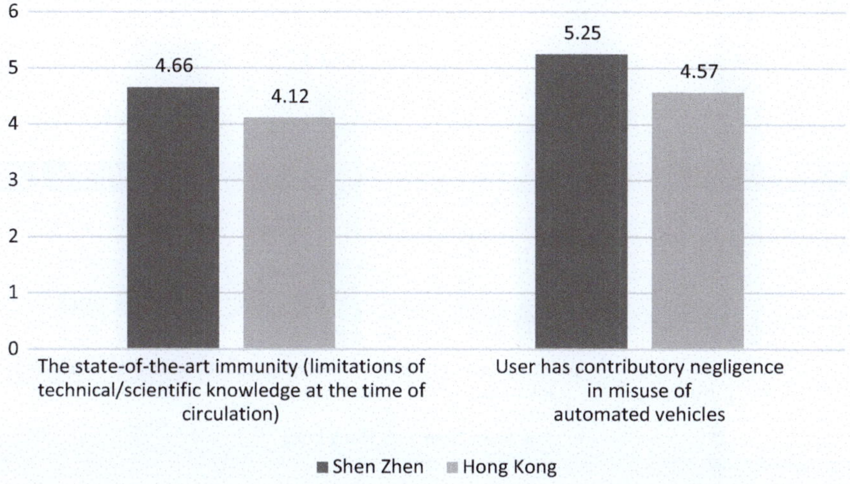

Fig. 9.20 How agreeable you are about the following reasons for liability exemption or reduction related to AVs?

Question 5: How agreeable are you to the following models of liability insurance for automated vehicle accidents?

Finding (Fig. 9.21): The scores were measured on a scale from 1 to 7, where 1 represents "total disagreed," 2 represents "very disagreed," 3 represents "slightly disagreed," 4 represents "neutral," 5 represents "slightly agreed," 6 represents "very agreed," and 7 represents "total agreed." Regarding the preference for optional liability insurance models, respondents from Shenzhen consistently demonstrated higher levels of agreement across all insurance models compared to those from Hong Kong, indicating a stronger preference or acceptance for robust insurance mechanisms in the context of AVs. Specifically, "the user's third-party liability insurance" showed the greatest difference, with Shenzhen scoring an average of 5.70 compared to Hong Kong's 5.15, indicating a variation of 0.55 ($p < 0.001$). This was followed by "the manufacturers' product liability insurance" (mean difference = 0.35, $p < 0.01$), "the industrial collective liability insurance" (mean difference = 0.31, $p < 0.01$), and "the operators' liability insurance" (mean difference = 0.33, $p < 0.01$). These statistically significant differences underscore a more pronounced support for comprehensive liability insurance frameworks in Shenzhen.

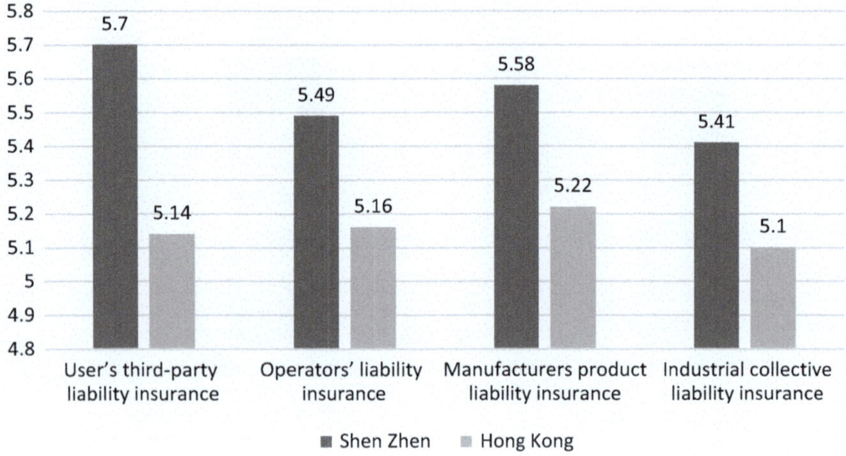

Fig. 9.21 How agreeable you are about the following models of liability insurance for automated vehicle accidents?

Among respondents from Shenzhen, the highest level of agreement was observed for "the user's third-party liability insurance" (M = 5.70), indicating a strong focus on protecting third parties. The lowest level of agreement was seen for "the industrial collective liability insurance" (M = 5.41), although it still received significant support. In contrast, in Hong Kong, the highest level of agreement was for "the manufacturers' product liability insurance" (M = 5.22), highlighting a specific concern for manufacturer accountability in case of product failure. The lowest level of agreement in Hong Kong was for "the industrial collective liability insurance" (M = 5.10), indicating a lack of consensus. Importantly, the differences between the highest and lowest agreement levels in Hong Kong were relatively small, suggesting a more consistent viewpoint across various insurance models compared to Shenzhen.

Comments on Legal Findings

In this section, we will delve deeper into the findings of the five survey questions that address (1) safety, (2) data privacy and security, (3) potentially liable parties, (4) exemptions from liability, and (5) models of liability insurance in the utilization of AVs. Through a comparative analysis of the survey results from Shenzhen and Hong Kong, we aim to derive significant and practical insights for stakeholders in formulating pertinent policies and regulations related to the development of AVs in these two cities.

Safety of (Highly) Autonomous Vehicles

The primary legal concern for potential users of autonomous vehicles (particularly level 4 and level 5 highly autonomous vehicles) is safety. Autonomous vehicles may pose risks to both the occupants inside the vehicle and other road users for two main reasons: (1) performance issues in typical scenarios, such as equipment or system failures, subpar system performance in challenging weather conditions, and interactions with other road users (including pedestrians, cyclists, and non-autonomous vehicles); and (2) performance challenges in unforeseen circumstances.

Unsurprisingly, the survey results show that both Shenzhen and Hong Kong rank highest in safety concerns regarding highly autonomous vehicles, whether in terms of car performance in typical or unexpected situations. To instill confidence in potential users of highly autonomous vehicles, it is crucial for these two cities to address how their laws can establish legal standards to ensure the safety control of AVs, road safety regulations, liability rules for AV accidents, and insurance mechanisms for liability. These regulations encompass criminal law, administrative law, and civil law, necessitating lawmakers to take a comprehensive approach to designing AV safety regulations.

However, respondents from Shenzhen show significantly greater concern about safety issues compared to their counterparts in Hong Kong, especially regarding car performance in typical scenarios. It is interesting to note that while Shenzhen participants exhibit a notably higher level of interest in Level 1 to Level 3 autonomous vehicles and generally display stronger interest across all levels of AVs compared to Hong Kong participants, they are also more vigilant about the potential risks associated with the unsafe performance of highly autonomous vehicles. This suggests that Shenzhen is poised to become a more enthusiastic market for AVs than Hong Kong once the AV industry can establish public confidence in the safety of these vehicles. Simultaneously, the government should implement stricter regulations for AV safety to nurture and enhance the AV market in Shenzhen. Given that Shenzhen can be seen as a representative city of mainland China, these observations may hold true for a larger market in mainland China.

The results of Question 1 further revealed that participants from Shenzhen were more concerned about the costs of learning to use self-driving vehicles and the availability of manual control in autonomous vehicles if the driver enjoys driving, compared to their counterparts in Hong Kong. This suggests that the industry and government in Shenzhen should make reasonable investments in public education on how to safely use autonomous cars to develop a new AV market. However, this does not imply that similar efforts should not be made in Hong Kong. Given the high level of concern shown by Shenzhen participants regarding the option of manual control for drivers who enjoy driving, the AV industry is suggested to consider this demand when designing and marketing highly autonomous vehicles in the future.

It is interesting to note that respondents from both Shenzhen and Hong Kong show relatively low concern about how highly autonomous vehicles will interact with other road users (such as pedestrians, cyclists, and non-autonomous cars) compared to their concerns about the performance of the vehicle in typical and unexpected situations. This suggests that while the ethical debates surrounding AV interactions with other road users may be controversial, they might not be a significant deterrent for people to use highly autonomous vehicles.

Data Privacy and Security

Unlike traditional vehicles, autonomous vehicles rely on big data for operation. This process involves collecting various data through sensors, integrating, analyzing, and interpreting the collected data using software algorithms, and utilizing the processed data to enable the AV to make immediate decisions such as determining driving routes, steering directions, braking, and speed adjustments, which are then executed by the AV control system. The second legal aspect concerning the use of AVs is data privacy and security.

Our survey indicates that participants from Shenzhen express statistically higher levels of concern across all surveyed dimensions compared to those from Hong Kong. Moreover, compared to participants in Hong Kong, Shenzhen participants generally showed a higher level of optimism for the benefits of AVs in terms of road safety, traffic efficiency and environmental protection; they are also more aware of the practical risk of data breaches associated with the use of AVs. Despite the common understanding that Hong Kong established personal data laws earlier than mainland China, our survey reveals that citizens' apprehensions regarding personal data protection in relation to the use of autonomous vehicles may not necessarily align with the development of personal data laws. One possible explanation is that citizens may be more vigilant about data breaches when personal data laws are less developed.

Above all, survey participants emphasize the threat of hacker attacks on vehicles and system security. Therefore, it is crucial for the laws in both Shenzhen and Hong Kong to prioritize safeguarding the data security of AVs by proactively preventing security threats and devising effective contingency plans for emergencies. The Chinese government has been actively promoting intelligent connected vehicles that facilitate extensive data sharing and communication with both internal and external devices. Hacker attacks targeting vehicles and AV operating systems can pose significant risks to vehicle safety, public safety, and even national security.

Next, our survey reveals that participants from Shenzhen have significant concerns about personal data privacy related to the use of autonomous vehicles. Specifically, issues such as "location and destination tracking" and "cross-border transfer and utilization of autonomous vehicle-generated big data" are of particular concern. This indicates that policymakers in Shenzhen need to implement adequate measures to safeguard the personal data of autonomous vehicle users, especially concerning their location tracking and the cross-border flow of data. Considering the potential value of AV-generated big data, regulations should also establish a transparent and robust legal framework governing the secondary use of AV-generated big data for public or commercial objectives.

As a precaution, while participants from Hong Kong may appear less sensitive to the risks of personal data security and privacy when using autonomous vehicles, this does not mean that Hong Kong lawmakers can afford to relax the protection of personal data for autonomous vehicle users. The existing personal data laws in Hong Kong have already established comprehensive legal principles and requirements that the autonomous vehicle industry must comply with. The pressing task for lawmakers is to delineate specific rules regarding personal data to align with the unique context of autonomous vehicles.

Potentially Liable Parties for AV Accidents

Survey question 3 outlines a range of potentially responsible parties in the event of car accidents involving AVs, especially driverless autonomous cars. Both Shenzhen and Hong Kong respondents generally agreed that the AV industry should bear primary strict liability for damages resulting from AV accidents, with the government potentially acting as a secondary party to provide reimbursement through a no-fault scheme. Sellers and owners of AVs are considered to be the least culpable. Additionally, there is a consensus that AV users should bear responsibility for damages if their misuse causes an accident.

The findings provide valuable insights on how to establish a foundational legal framework for AV liability laws for policymakers in both cities. In cases where accidents are caused by factors such as defective design, manufacturing flaws, programming errors, inadequate warnings and instructions, or human errors, it triggers product and service liability of manufacturers, programmers, and operators.

Chinese national product liability law, which applies to Shenzhen, follows a strict liability approach. This means that manufacturers are held liable for compensating damages caused by defective products regardless of whether they took reasonable care to prevent the associated risks. On the other hand, Hong Kong continues to adhere to a fault-based product liability law, where consumers must prove that manufacturers are at fault for the defect in question. This divergence in the two legal systems may explain why respondents from Shenzhen are more inclined to support the strict liability of AV manufacturers, programmers, and operators for accidents caused by defective AVs than respondents from Hong Kong.

However, when compared to the majority of common law systems that have already embraced a strict liability framework, Hong Kong's product liability law seems conservative and outdated. In terms of the parties that could be held responsible for AV accidents, respondents from Hong Kong have expressed support for a strict liability approach. This suggests an opening for potential reform of Hong Kong's product liability laws, particularly concerning defective autonomous vehicles. Furthermore, considering that traditional vehicles have been permitted to cross the border between Shenzhen and Hong Kong, if this policy is extended to AVs in the future, it would be desirable for both legal systems to adopt a harmonized approach in dealing with product liability issues related to cross-border AVs.

Lastly, while existing Chinese law holds manufacturers and sellers jointly liable for defective products, respondents in Shenzhen exhibit less enthusiasm for imposing joint liability on sellers for defective AVs. This could be attributed to the fact that, unlike consumable goods, the identity and details of AV manufacturers are readily available to users, making it relatively easy for consumers to identify and reach out to AV manufacturers in the event of disputes.

Exemptions from Liability

The rules governing exemptions from liability also determine the liability extent of potentially responsible parties in AV accidents. Our survey did not reveal a statistically significant difference on this issue between respondents from Shenzhen and Hong Kong. They generally agreed that state-of-the-art immunity and the user's contributory negligence in misusing AVs should be factors that reduce the liability of parties to pay compensation for damages resulting from AV accidents. This suggests that users understand the limitations of technical and scientific knowledge at the time of circulation and are willing to accept accidents resulting from these limitations.

However, if the law recognizes state-of-the-art immunity, it must address two other related issues: (1) Should AV manufacturers have a duty to rectify defects that were not detected due to limitations in technical and scientific knowledge at the time of circulation once new knowledge becomes available? (2) Should victims be provided with alternative compensation schemes (such as social relief, insurance, or a no-fault scheme) if their claims are unsuccessful due to state-of-the-art immunity? These issues are awaiting clear answers from lawmakers in both cities.

Liability Insurance Models

When Shenzhen respondents show a greater interest in AVs with varying levels of automation compared to their counterparts in Hong Kong, they also pay close attention to the liability insurance mechanisms for AV accidents. They express the highest level of concern regarding the user's third-party liability insurance, indicating, as AV users, their greatest worry about potential legal liability towards third-party victims in AV accidents. The extent to which the legal system can offer them adequate legal solutions may influence their willingness to use AVs and impact the growth of the

AV market. This underscores the importance of Shenzhen lawmakers prioritizing this issue when developing liability insurance frameworks for AVs. In contrast, respondents from Hong Kong prioritize manufacturers' product liability insurance as the most critical concern, despite a slight variance in agreement levels across different types of liability insurance models.

Simultaneously, the survey findings on Question 5 indicate a need for a comprehensive liability insurance framework for AVs and underscore the crucial role that liability insurance mechanisms should play in allocating and distributing the risks and expenses associated with AV accidents, as well as alleviating users' concerns about the uncertain legal risks of compensation. This comprehensive liability insurance framework should engage various stakeholders, such as users, manufacturers, operators, and the entire AV industry. It is imperative for lawmakers in both cities to make an effort to establish a reliable and robust liability insurance framework to bolster consumer trust in the use of AVs.

Concluding Remarks

Overall, survey respondents in both Shenzhen and Hong Kong exhibited enthusiasm and interest in AVs, with a stronger interest observed in higher levels of automation. Notable differences in interest were noted for AVs ranging from Levels 1 to 3, but this discrepancy was not evident for AVs at Levels 4 to 5. Participants from both regions expressed interest in driving or riding in AVs and displayed confidence that AVs would bring benefits to both users and society. While respondents from Shenzhen generally held a more optimistic view towards AVs compared to their counterparts in Hong Kong, our findings suggest that all respondents from these two markets in Greater China are supportive of the development and utilization of AVs in the future. The key highlights from our survey findings provide valuable insights for the industry, lawmakers, and policymakers in both mainland China and Hong Kong on how to design AVs and prioritize regulatory and policy objectives for future AV usage.

Our survey on the ethical implications of AVs revealed some intriguing contrasts. Firstly, unlike survey results in Western countries, Chinese respondents did not view AVs programmed to prioritize the safety of external road users over internal users in the event of brake failure or other emergencies as a deterrent to purchasing or using such AVs. In essence, they were open to the idea of AVs being ethically designed to prioritize greater moral goals, even if it meant sacrificing the interests of internal users. Secondly, respondents from Shenzhen and Hong Kong, despite both being influenced by Confucian familism, displayed differing perspectives when external road users' well-being was considered more important than that of internal users. Hong Kong, having been influenced by British utilitarianism, tended to agree that increasing total pleasures and reducing total pains may be necessary, regardless of whether their family members are involved. However, both groups concurred on prioritizing the lives of multiple pedestrians over a small number of users inside AVs. Thirdly, respondents from both cities expressed a desire to retain a steering wheel in AVs to retain control and make human decisions that align with their own ethical values during a traffic crisis.

The findings from our survey regarding the legal perspective reveal that concerns about safety and data privacy and security are the most significant worries among respondents regarding highly autonomous vehicles. Participants who expressed a higher level of interest in AVs also demonstrated greater vigilance towards safety concerns. Similarly, the same trend was observed in terms of data privacy and security concerns, regardless of the duration for which personal data protection measures have been in place. When discussing potential liable parties in AV accidents, respondents from both Shenzhen and Hong Kong agreed that the AV industry should bear primary responsibility and strict liability for damages resulting from traffic accidents, with the government potentially serving as a secondary party to assist injured parties through a no-fault system. Sellers and owners of AVs were considered to be the least culpable and AV users should be held accountable for damages if their misuse leads to an accident. Furthermore, there was also a consensus that state-of-the-art immunity and the user's contributory negligence in misusing AVs should be factors that reduce the liability of parties to compensate for AV accidents damages. In terms of liability insurance models, respondents from Shenzhen showed a greater concern for AV users' third-party liability insurance, while those from Hong Kong emphasized the importance of manufacturers' product liability insurance. Nevertheless, all survey participants concurred on the necessity of a comprehensive liability insurance framework for AVs, as a reliable and robust system is crucial to enhance consumer trust in the use of AVs.

References

Althaus, S. L. (2003). *Collective preferences in democratic politics: Opinion surveys and the will of the people*. Cambridge University Press.

Awad, E., Dsouza, S., Kim, R., et al. (2018). The moral machine experiment. *Nature, 563*, 59–64.

Bell, D. A., & Li, C. (2013). *The East Asian challenge for democracy: Political meritocracy in comparative perspective*. Cambridge University.

Bennefon, J., Shariff, A., & Rahwan, I. (2016). The social dilemma of autonomous vehicles. *Science, 352*, 1573–1576.

Carlsson, F., Daruvala, D., & Jaldell, H. (2010). Preferences for lives, injuries, and age: A stated preference survey. *Accident; Analysis and Prevention, 42*, 1814–1821.

Cruz, C. J. P., Matus, K., & Gietel-Basten, S. (2023). The extent of use of surveys in policymaking: The case of Hong Kong. *Evidence & Policy*, 1–22.

Nabatchi, T., & Amsler, L. B. (2014). Direct public engagement in local government. *The American Review of Public Administration, 44*(4_suppl), 63S–88S.

Othman, K. (2021). Public acceptance and perception of autonomous vehicles: A comprehensive review. *AI & Ethics, 1*, 355–387.

Singer, P. (1972). Famine, affluence, and morality. *Philosophy & Public Affairs, 1*(3), 229–243.

Wong, D. (2023). Chinese ethics. In *Stanford encyclopedia of philosophy*. https://plato.stanford.edu/entries/ethics-chinese/#Aca

Index[1]

[1] Note: Page numbers followed by 'n' refer to notes.

C. Ding, R. Fan (eds.), *Chinese Ethical and Legal Perspectives on Automated Vehicles*,
https://doi.org/10.1007/978-981-96-1277-2

The manufacturer's authorised representative in the EU is Springer Nature Customer Service Centre GmbH, Europaplatz 3, 69115 Heidelberg, Germany. If you have any concerns regarding our products, please contact ProductSafety@springernature.com

Printed and bound by CPI Group (UK) Ltd, Croydon, CR0 4YY
01/10/2026
02225927-0001